I0064909

Introduction to Glass Physics

Introduction to Glass Physics

Fiona Allen

WILLFORD PRESS
www.willfordpress.com

Published by Willford Press,
118-35 Queens Blvd., Suite 400,
Forest Hills, NY 11375, USA

Copyright © 2022 Willford Press

This book contains information obtained from authentic and highly regarded sources. All chapters are published with permission under the Creative Commons Attribution Share Alike License or equivalent. A wide variety of references are listed. Permissions and sources are indicated; for detailed attributions, please refer to the permissions page. Reasonable efforts have been made to publish reliable data and information, but the authors, editors and publisher cannot assume any responsibility for the validity of all materials or the consequences of their use.

Trademark Notice: Registered trademark of products or corporate names are used only for explanation and identification without intent to infringe.

ISBN: 978-1-64728-358-2

Cataloging-in-Publication Data

Introduction to glass physics / Fiona Allen.
p. cm.
Includes bibliographical references and index.
ISBN 978-1-64728-358-2
1. Glass. 2. Physics. 3. Optics. I. Allen, Fiona.
TP857 .I58 2022
666.1--dc23

For information on all Willford Press publications
visit our website at www.willfordpress.com

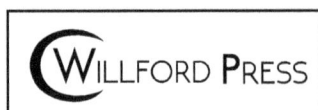

WILLFORD PRESS

TABLE OF CONTENTS

Preface VII

Chapter 1   **What is Glass Physics?** 1

- Glass   1
- Industrial Glass and its Properties   4

Chapter 2   **Types of Glass** 18

- Safety Glass   18
- Smart Glass   18
- Fused Quartz   23
- Photosensitive Glass   29
- Foturan   32
- Cer-Vit   37
- Porous Glass   37
- Temperature Sensitive Glass   41
- Tempered Glass   42
- Beveled Glass   46
- Hebron Glass   47
- Soda-lime Glass   49
- Heatable Glass   49
- Plate Glass   54
- Borosilicate Glass   55
- Tiffany Glass   55
- Stained Glass   60

Chapter 3   **Processes of Glass Preparation** 70

- Glass Melting   70
- Glass Batch Calculation   70
- Calculation of Glass Properties   72
- Glass Coating   76

- Annealing of Glass   78
- Glass Inspection   79

Chapter 4   **Glass Production Techniques**.. **90**

- Commercial Techniques   90
- Artistic Techniques   108

Chapter 5   **Applications** **123**

- Medical Applications of Glass   123
- Industrial Applications of Glass   158

Permissions

Index

PREFACE

It is with great pleasure that I present this book. It has been carefully written after numerous discussions with my peers and other practitioners of the field. I would like to take this opportunity to thank my family and friends who have been extremely supporting at every step in my life.

Glass is a transparent amorphous solid which is non-crystalline. It has a wide range of uses including practical, decorative and technological applications. The field of physics which studies characteristics and properties of glass is known as glass physics. Glass transition is the reversible and gradual transition in amorphous liquids from a brittle glassy stage into viscous stage. Its transition takes place when the temperature is increased. The temperature at which it takes place is called its glass transition temperature. Glass defects are central to glass physics. It is broadly applicable to domains ranging from computer science to biophysics and soft condensed-matter physics. This book provides comprehensive insights into the field of glass physics. Also included herein is a detailed explanation of the various concepts and applications of this field. Through this book, we attempt to further enlighten the readers about the new concepts of glass physics.

The chapters below are organized to facilitate a comprehensive understanding of the subject:

Chapter – What is Glass Physics?

Glass is the inorganic hard, brittle substance that is usually transparent or translucent which is made by fusing soda and lime and cooling rapidly. Industrial glass is a solid material that has three main properties of lustre, transparency and durability. This is an introductory chapter which will introduce briefly all these significant aspects of glass as well as their uses.

Chapter – Types of Glass

There are various types of glasses used in various purposes of construction and engineering. A few types of such glasses are safety glass, fused quartz, hebron glass, soda lime glass, plate glass, tiffany glass, stained glass, etc. This chapter has been carefully written to provide an easy understanding of these varied types of glass.

Chapter – Processes of Glass Preparation

The processing of glass consists of various important manufacturing steps such as glass melting, glass batch calculation, calculation of glass properties, glass coating, annealing of glass and glass inspection. This chapter closely examines these key stages of glass manufacturing to provide an extensive understanding of the subject.

Chapter – Glass Production Techniques

The glass production techniques can be classified into commercial techniques and artistic techniques. Commercial techniques include glassblowing, glass casting, vitrification, etc. and artistic techniques include caneworking, glass fusing, slumping, etc. All these diverse glass production techniques have been carefully analyzed in this chapter.

Chapter – Applications

Glass has a wide range of applications in different fields of medicine and industry. A few of its medical applications are usage of glass in beaker, smartglasses, bioglass, bioactive glass, etc. and its industrial applications include the use of glass in mirror, windshield, liquid crystal display, etc. The topics elaborated in this chapter will help in gaining a better perspective about these applications of glass.

Fiona Allen

CHAPTER 1

What is Glass Physics?

Glass is the inorganic hard, brittle substance that is usually transparent or translucent which is made by fusing soda and lime and cooling rapidly. Industrial glass is a solid material that has three main properties of lustre, transparency and durability. This is an introductory chapter which will introduce briefly all these significant aspects of glass as well as their uses.

GLASS

Glass is an inorganic solid material that is usually transparent or translucent as well as hard, brittle, and impervious to the natural elements. Glass has been made into practical and decorative objects since ancient times, and it is still very important in applications as disparate as building construction, housewares, and telecommunications. It is made by cooling molten ingredients such as silica sand with sufficient rapidity to prevent the formation of visible crystals.

Louvre Museum, Paris, with steel-and-glass pyramid.

Stained glass and the aesthetic aspects of glass design are described in stained glass and glassware. The composition, properties, and industrial production of glass are covered in industrial glass. The physical and atomic characteristics of glass are treated in amorphous solid.

The varieties of glass differ widely in chemical composition and in physical qualities. Most varieties, however, have certain qualities in common. They pass through a viscous stage in cooling from a state of fluidity; they develop effects of colour when the glass mixtures are fused with certain metallic oxides; they are, when cold, poor conductors both of electricity and of heat; most types are easily fractured by a blow or shock and show a conchoidal fracture; and they are but slightly affected by ordinary solvents but are readily attacked by hydrofluoric acid.

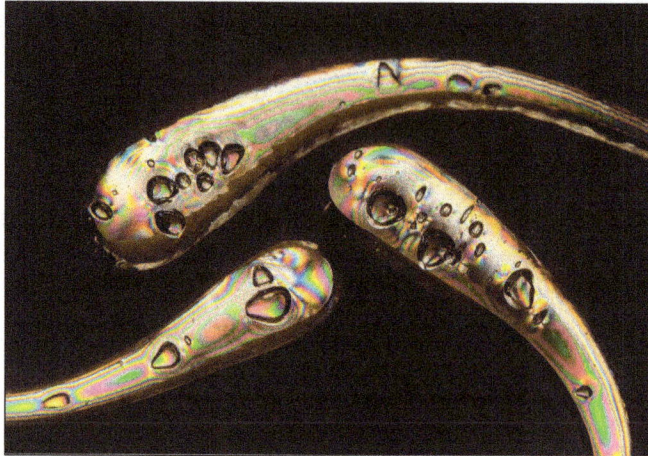

Prince Rupert's Drop.

The Prince Rupert's Drop is a droplet of glass formed by the rapid cooling of molten glass in cold water. A novelty in the 1600s, the droplets are used today to demonstrate the strength of tempered glass. The image here, produced with the use of polarized lenses, shows stress and potential energy stored in the glass as a rainbow.

Commercial Glass Composition

Commercial glasses may be divided into soda–lime–silica glasses and special glasses, most of the tonnage produced being of the former class. Such glasses are made from three main materials—sand (silicon dioxide, or SiO_2), limestone (calcium carbonate, or $CaCO_3$), and sodium carbonate (Na_2CO_3). Fused silica itself is an excellent glass, but, as the melting point of sand (crystalline silica) is above 1,700 °C (3,092 °F) and as it is very expensive to attain such high temperatures, its uses are restricted to those in which its superior properties—chemical inertness and the ability to withstand sudden changes of temperature—are so important that the cost is justified. Nevertheless, the production of fused silica glass is quite a large industry; it is manufactured in various qualities, and, when intended for optical purposes, the raw material used is rock crystal rather than quartz sand.

To reduce the melting point of silica, it is necessary to add a flux; this is the purpose of the sodium carbonate (soda ash), which makes available the fluxing agent sodium oxide. By adding about 25 percent of the sodium oxide to silica, the melting point is reduced from 1,723 to 850 °C (3,133 to 1,562 °F). But such glasses are easily soluble in water (their solutions are called water glass). The addition of lime (calcium oxide, or CaO), supplied by the limestone, renders the glass insoluble again, but too much makes a glass prone to devitrification—i.e., the precipitation of crystalline phases in certain ranges of temperature. The optimum composition is about 75 percent silica, 10 percent lime, and 15 percent soda, but even this is too liable to devitrification during certain mechanical forming operations to be satisfactory.

In making sheet glass it is customary to use 6 percent of lime and 4 percent of magnesia (magnesium oxide, or MgO), and in bottle glass about 2 percent alumina (aluminum oxide, or Al_2O_3) is often present. Other materials are also added, some being put in to assist in refining the glass (i.e., to remove the bubbles left behind in the melting process), while others are added to improve its colour. For example, sand always contains iron as an impurity, and, although the material used for making bottles is specially selected for its low iron content, the small traces of impurity still impart an undesirable green colour to the container; by the use of selenium and cobalt oxide together with traces of arsenic trioxide and sodium nitrate, it is possible to neutralize the green colour and produce a so-called white (decolourized) glass.

Optical and High-temperature Glass

Glasses of very different, and often much more expensive, compositions are made when special physical and chemical properties are necessary. For example, in optical glasses, a wide range of compositions is required to obtain the variety of refractive index and dispersion needed if the lens designer is to produce multicomponent lenses that are free from the various faults associated with a single lens, such as chromatic aberration. High-purity, ultratransparent oxide glasses have been developed for use in fibre-optic telecommunications systems, in which messages are transmitted as light pulses over glass fibres.

When ordinary glass is subjected to a sudden change of temperature, stresses are produced in it that render it liable to fracture; by reducing its coefficient of thermal expansion, however, it is possible to make it much less susceptible to thermal shock. The glass with the lowest expansion coefficient is fused silica. Another well-known example is the borosilicate glass used for making domestic cookware, which has an expansion coefficient only one-third that of the typical soda–lime–silica glass. In order to effect this reduction, much of the sodium oxide added as a flux is replaced by boric oxide (B_2O_3) and some of the lime by alumina. Another familiar special glass is the lead crystal glass used in the manufacture of superior tableware; by using lead monoxide (PbO) as a flux, it is possible to obtain a glass with a high refractive index and, consequently, the desired sparkle and brilliance.

Adding Colour and Special Properties

The agents used to colour glass are generally metallic oxides. The same oxide may produce different colours with different glass mixtures, and different oxides of the same metal may produce different colours. The purple-blue of cobalt, the chrome green or yellow of chromium, the dichroic canary colour of uranium, and the violet of manganese are constant. Ferrous oxide produces an olive green or a pale blue according to the glass with which it is mixed. Ferric oxide gives a yellow colour but requires an oxidizing agent to prevent reduction to the ferrous state. Lead gives a pale yellow colour. Silver oxide gives a permanent yellow stain. Finely divided vegetable charcoal added to a soda–lime glass gives a yellow colour. Selenites and selenates give a pale pink or pinkish yellow. Tellurium appears to give a pale pink tint. Nickel with a potash–lead glass gives a violet colour, and a brown colour with a soda–lime glass. Copper gives a peacock blue, which becomes green if the proportion of the copper oxide is increased.

Wine goblet, blue glass decorated with white and gold enamel.

An important class of materials is the chalcogenide glasses, which are selenides, containing thallium, arsenic, tellurium, and antimony in various proportions. They behave as amorphous semiconductors. Their photoconductive properties are also valuable.

Certain metallic glasses have magnetic properties; their characteristics of ease of manufacture, magnetic softness, and high electrical resistivity make them useful in the magnetic cores of electrical power transformers.

INDUSTRIAL GLASS AND ITS PROPERTIES

Industrial Glass, also called as architectural glass is a solid material that is normally lustrous and transparent in appearance and that shows great durability under exposure to the natural elements. These three properties—lustre, transparency, and durability—make glass a favoured material for such household objects as windowpanes, bottles, and lightbulbs. However, neither any of these properties alone nor all of them together are sufficient or even necessary for a complete description of glass. Defined according to modern scientific beliefs, glass is a solid material that has the atomic structure of a liquid.

Normally, glass is formed upon the cooling of a molten liquid in such a manner that the ordering of atoms into a crystalline formation is prevented. Instead of the abrupt change in structure that takes place in a crystalline material such as metal as it is cooled below its melting point, in the cooling of a glass-forming liquid there is a continuous stiffening of the fluid until the atoms are virtually frozen into a more or less random arrangement similar to the arrangement that they had in the fluid state. Conversely, upon application of heat to solid glass, there is a gradual softening of the structure until it reaches the fluid state. This monotonically changing property, known as viscosity, enables glass products to be made in a continuous fashion, with raw materials melted to a homogeneous liquid, delivered as a viscous mass to a forming machine to make a specific product, and then cooled to a hard and rigid condition.

Glass Compositions and Applications

Silica-based

Of the various glass families of commercial interest, most are based on silica, or silicon dioxide (SiO_2), a mineral that is found in great abundance in nature—particularly in quartz and beach sands. Glass made exclusively of silica is known as silica glass, or vitreous silica. (It is also called fused quartz if derived from the melting of quartz crystals.) Silica glass is used where high service temperature, very high thermal shock resistance, high chemical durability, very low electrical conductivity, and good ultraviolet transparency are desired. However, for most glass products, such as containers, windows, and lightbulbs, the primary criteria are low cost and good durability, and the glasses that best meet these criteria are based on the soda-lime-silica system. Examples of these glasses are shown in the table Composition of representative oxide glasses.

Table: Composition of representative oxide glasses.

Glass family	Glass application	Oxide ingredient (percent by weight)				
		Silica (SiO_2)	Soda (Na_2O)	Lime (CaO)	Alumina (Al_2O_3)	Magnesia (MgO)
Vitreous silica	Furnace tubes, silicon melting crucibles	100.0	—	—	—	—
Soda-lime silicate	Window	72.0	14.2	10.0	0.6	2.5
	Container	74.0	15.3	5.4	1.0	3.7
	Bulb and tube	73.3	16.0	5.2	1.3	3.5

Glass family	Glass application					
	Tableware	74.0	18.0	7.5	0.5	—
Sodium borosilicate	Chemical glassware	81.0	4.5	—	2.0	—
Lead-alkali silicate	Lead "crystal"	59.0	2.0	—	0.4	—
	Television funnel	54.0	6.0	3.0	2.0	2.0
Aluminosilicate	Glass halogen lamp	57.0	0.01	10.0	16.0	7.0
	Fibreglass "E"	52.9	—	17.4	14.5	4.4
Optical	"Crown"	68.9	8.8	—	—	—

Glass family	Glass application	Oxide ingredient (percent by weight)				
		Boron oxide (B_2O_3)	Barium oxide (BaO)	Lead oxide (PbO)	Potassium oxide (K_2O)	Zinc oxide (ZnO)
Vitreous silica	Furnace tubes, silicon melting crucibles	—	—	—	—	—
Soda-lime silicate	Window	—	—	—	—	—
	Container	—	trace	—	0.6	—
	Bulb and tube	—	—	—	0.6	—
	Tableware	—	—	—	—	—
Sodium borosilicate	Chemical glassware	12.0	—	—	—	—
Lead-alkali silicate	Lead "crystal"	—	—	25.0	12.0	1.5
	Television funnel	—	—	23.0	8.0	—
Aluminosilicate	Glass halogen lamp	4.0	6.0	—	trace	—
	Fibreglass "E"	9.2	—	—	1.0	—
Optical	"Crown"	10.1	2.8	—	8.4	1.0

After silica, the many "soda-lime" glasses have as their primary constituents soda, or sodium oxide (Na_2O; usually derived from sodium carbonate, or soda ash), and lime, or calcium oxide (CaO; commonly derived from roasted limestone). To this basic formula other ingredients may be added in order to obtain varying properties. For instance, by adding sodium fluoride or calcium fluoride, a translucent but not transparent product known as opal glass can be obtained. Another silica-based variation is borosilicate glass, which is used where high thermal shock resistance and high chemical durability are desired—as in chemical glassware and automobile headlamps. In the past, leaded "crystal" tableware was made of glass containing high amounts of lead oxide (PbO), which imparted to the product a high refractive index (hence the brilliance), a high elastic modulus (hence the sonority, or "ring"), and a long working range of temperatures. Lead oxide is also a major component in glass solders or in sealing glasses with low firing temperatures.

Other silica-based glasses are the aluminosilicate glasses, which are intermediate between vitreous silica and the more common soda-lime-silica glasses in thermal properties as well as cost; glass fibres such as E glass and S glass, used in fibre-reinforced plastics and in thermal-insulation wool; and optical glasses containing a multitude of additional major constituents.

Nonsilica

Oxide glasses not based on silica are of little commercial importance. They are generally phosphates and borates, which have some use in bioresorbable products such as surgical mesh and time-release capsules.

Non Oxide Glasses

Heavy-metal Fluoride Glasses

Of the non oxide glasses, the heavy-metal fluoride glasses (HMFGs) have potential use in telecommunications fibres, owing to their relatively low optical losses. However, they are also extremely difficult to form and have poor chemical durability. The most studied HMFG is the so-called ZBLAN group, containing fluorides of zirconium, barium, lanthanum, aluminum, and sodium.

Glassy Metals

Another non oxide group is the glassy metals, formed by high-speed quenching of fluid metals. Perhaps the most studied glassy metal is a compound of iron, nickel, phosphorus, and boron that is commercially available as Metglas (trademark). It is used in flexible magnetic shielding and power transformers.

Semiconducting Solids

A final class of non oxide, noncrystalline substances is the chalcogenides, which are formed by melting together the chalcogen elements sulfur, selenium, or tellurium with elements from group V (e.g., arsenic, antimony) and group IV (e.g., germanium) of the periodic table. Owing to their semiconducting properties, chalcogenides have found use in threshold and memory switching devices and in xerography. A related end-member of this group is the elemental amorphous semiconductor solids, such as amorphous silicon (a-Si) and amorphous germanium (a-Ge). These materials are the basis of most photovoltaic applications, such as the solar cells in pocket calculators. Amorphous solids have a liquidlike atomic order but are not considered to be true glasses because they do not exhibit a continuous transformation into the liquid state upon heating.

Glass Ceramics

In some glasses it is possible to bring about a certain degree of crystallization in the normally random atomic structure. Glassy materials that exhibit such a structure are called glass ceramics. Commercially useful glass ceramics are those in which a high density of uniformly sized, nonoriented crystals has been achieved through the bulk of the material, rather than at the surface or in discrete regions. Such products invariably possess strengths far exceeding those of the parent glass or of the corresponding ceramic. Outstanding examples are Corning Ware (trademark) cooking vessels and Dicor (trademark) dental implants.

Glass Composites

In addition to the glass ceramics, useful products of glass may be made by mixing ceramic, metal, and polymer powders. Most products made from such blends, or composites, exhibit properties that are combinations of the properties of the various ingredients. Good examples of composite products are glass-fibre reinforced plastics, for use as tough elastic solids, and thick-film conductor, resistor, and dielectric pastes with tailored electrical properties for the packaging of microcircuits.

Natural Glasses

Several inorganic glasses are found in nature. These include obsidians (volcanic glasses), fulgarites (formed by lightning strikes), tektites found on land in Australasia and associated microtektites from the bottom of the Indian Ocean, moldavites from central Europe, and Libyan Desert glass from western Egypt. Owing to their extremely high chemical durability under the sea, microtektite compositions are of significant commercial interest for hazardous waste immobilization or conversion.

Properties of Glass

At ordinary temperatures, glass is a nearly perfect elastic solid, an excellent thermal and electrical insulator, and very resistant to many corrosive media. (Its optical properties, however, vary greatly, depending on the light wavelengths employed.) The more or less random order of atoms is ultimately responsible for many of the properties that distinguish glass from other solids. One unique attribute of special importance may be called the isotropicity of properties, meaning that such properties as tensile strength, electrical resistance, and thermal expansion are of equal magnitude in any direction through the material.

As a glass-forming melt is cooled through the transition range, its structure relaxes, or changes continuously, from that of a liquid to that of a solid. The properties of solid glass reflect the extent of this structural relaxation. Indeed, glass can be said to retain a memory of the temperature-time schedule through the transition. Evidence of this "thermal history" is wiped out only after the glass has been reheated to the liquid state.

Most properties of glass—except for elastic and strength behaviour in the solid state—are sensitive to its chemical composition and, hence, its atomic structure. In oxide glasses, the specific composition-structure-property relationships are based upon the following factors: (1) the coordination number of the network-forming (NWF) ion, (2) the connectivity of the structure, as determined by the concentration of nonbridging oxygens, which, in turn, is determined by the concentration and nature of network-modifying (NWM) ions, (3) the openness of the structure, determined, again, by the concentration of NWM ions, and (4) the mobility of the NWM ions. Thus, tetrahedrally connected networks, such as those formed by silicates and illustrated in, are more viscous than triangularly connected networks, such as those formed by borates. In silicates, the addition of network-modifying alkali ions would raise the concentration of nonbridging oxygens, and the resulting lowered connectivity would lead to a lowering of viscosity. Networks in which the interstitial spaces are less filled with NWM ions possess lower density and allow greater permeation of gases through them. Since alkali ions are the most mobile species through interstices of oxide glasses, the higher the alkali concentration, the lower the chemical durability and electrical resistivity of the material.

Because glass generally acts as if it were a solution, many of its properties can be estimated by applying what are known as additivity relationships over a narrow range of compositions. In additivity relationships, it is assumed that each ingredient in a glass contributes to the properties of the glass by an amount equal to the concentration of that ingredient multiplied by a specific additivity factor. Many properties of soda-lime-silica glasses follow such relationships closely.

Physical Properties

Density

In the random atomic order of a glassy solid, the atoms are packed less densely than in a corresponding crystal, leaving larger interstitial spaces, or holes between atoms. These interstitial spaces collectively make up what is known as free volume, and they are responsible for the lower density of a glass as opposed to a crystal. For example, the density of silica glass is about 2 percent lower than that of its closest crystalline counterpart, the silica mineral low-cristobalite. The addition of alkali and lime, however, would cause the density of the glass to increase steadily as the network-modifying sodium and calcium ions filled the interstitial spaces. Thus, commercial soda-lime-silica glasses have a density greater than that of low-cristobalite. Density follows additivity behaviour closely.

The densities of representative oxide glasses are shown in the table of properties of oxide glasses.

Table: Properties of oxide glasses.

Glass Family	Density (gm/cm³)	Maximum Service Temperature (°C)	Softening Point (°C)	Working Point (°C)	Linear Thermal-expansion Coefficient (per °C)
Vitreous silica	2.20	1,000–1,150	1,580–1,670	>2,000	5.5×10^{-7}
Soda-lime silicate	2.49	500	750	1,000	$85–95 \times 10^{-7}$
Sodium borosilicate	2.23	550	820	1,245	33×10^{-7}
Lead-alkali silicate	3.02	450	677	985	99×10^{-7}
Aluminosilicate	2.64	680	910	1,175	48×10^{-7}
Optical	2.51	550	719	941	71×10^{-7}

Glass Family	Weatherabiity (0 = poor, 4 = excellent)	Electrical Conductivity (mho/cm at 25 °C)	Dielectric Constant (at 1 MHz and 20 °C)	Refractive Index	
Vitreous silica	4	10^{-18}	3.8	1.459	
Soda-lime silicate	2	10^{-12}	7.0	1.51	
Sodium borosilicate	3.5	10^{-15}	5.1	1.474	
Lead-alkali silicate	2	$>10^{-17}$	6.7	1.563	
Aluminosilicate	4	$>10^{-17}$	6.3	1.547	
Optical	3.5	10^{-16}	6.5	1.517	

Elasticity and Plasticity

Because of the isotropic nature of glass, only two independent elastic moduli are normally measured: Young's modulus, which measures the ability of a solid to recover its original dimensions after being subjected to lengthwise tension or compression; and shear modulus, which measures its ability to recover from transverse stress. In oxide glasses, both Young's modulus and shear modulus do not strongly depend upon the chemical composition.

The hardness of glass is measured by a diamond microindenter. Application of this instrument to a glassy surface leaves clear evidence of plastic deformation—or a permanent change in dimension. Otherwise, plastic deformation of glass (or ductility), which is generally observed in strength tests as the necking of a specimen placed under tension, is not observed; instead, glass failure is brittle—that is, the glass object fractures suddenly and completely. This behaviour can be explained by the atomic structure of a glassy solid. Since the atoms in molten glass are essentially frozen in their amorphous order upon cooling, they do not orient themselves into the sheets or planes that are typical of growing crystalline grains. The absence of such a growth pattern means that no grain boundaries arise between planes of different orientation, and therefore there are no barriers that might prevent defects such as cracks from extending quickly through the material. The absence of dislocations causes glass not to display ductility, the property of yielding and bending like metal.

Strength and Fracturing

Glass is exceptionally strong, much stronger than most metals, when tested in the pristine state. Under pure compression, glass may undergo a more or less reversible compression but not fracture. Its theoretical strength in tension is estimated to be 14 to 35 gigapascals (2 to 5 million pounds per square inch); glass fibres produced under very careful drawing conditions have approached 11.5 gigapascals in strength. The strength of most commercial glass products, on the other hand, ranges between only 14 and 175 megapascals (2,000 and 25,000 pounds per square inch), owing to the presence of scratches and microscopic flaws, generally on the surface. Apparently, surface flaws are produced in glass by abrasion with most solids—even by the touch of a finger and particularly by another piece of glass that rubs against it during manufacture. Flaws have a stress-concentrating effect; that is, the effective stress at the tip of a flaw can be easily 100 to 1,000 times greater than that applied. Tensile stresses in excess of a low limit, called the fatigue limit, cause the flaw to undergo a subcritical crack growth. Eventually, depending on the applied stress, the shape of the flaw, the temperature, and even the corrosiveness of the environment, the growth velocity of the crack approaches its terminal limit, and failure becomes imminent. Thus, under a tensile loaded condition, all glass experiences static fatigue and eventually fails. The crack growth velocities are higher with higher magnitudes of tensile stress, sharper flaws (where the tip radius is much smaller than the length), higher temperatures, and higher humidity.

A glass fracture may be examined visually or with a (generally) low-power stereo microscope. Starting from its point of origin, the fracture front travels slowly, producing a nearly semicircular shiny surface called the mirror. The radius of the mirror is inversely related to the fracture stress and, hence, is indicative of the violence of the fracture. (For instance, a thermal fracture generally produces a large mirror, whereas a mechanical fracture often displays a small mirror.) The edges of the mirror have a fine fibrous or misty texture, called the mist. Surrounding the mist are wider and deeper radial ridges, with slivers of glass lifted out. Known as the hackle, these ridges ultimately lead to crack branching. Fracture travels faster in a region that is under tensile stress than in a region of compression; severe compression causes the direction of crack growth to twist, producing a twist hackle or river pattern. Penetration by a pointed object, such as a bullet, often produces what is known as a Hertzian cone fracture, in which an expanding cone of glass is ejected from the side of glass opposite to the impact.

Fractography of glass is important in manufacture and service, in that it is equivalent to a postmortem examination. An experienced fractographer can often pinpoint the origin, the cause, and the circumstances of product failure.

Thermal Properties

Viscosity

As can be seen from figure, the viscosity of glass, as measured in centimetre-gram-second units known as poise, decreases with rising temperature. Figure also indicates the temperatures at which certain glasses reach standard viscosity reference points that are important in glassmaking. For instance, the working point, the temperature at which a gob of molten glass may be delivered to a forming machine, is equivalent to the temperature at which viscosity is 10^4 poise. The softening point, at which the glass may slump under its own weight, is defined by a viscosity of $10^{7.65}$ poise, the annealing point by 10^{13} poise, and finally the strain point by $10^{14.5}$ poise. Upon further cooling, viscosity increases rapidly to well beyond 10^{18} poise, where it can no longer be measured meaningfully.

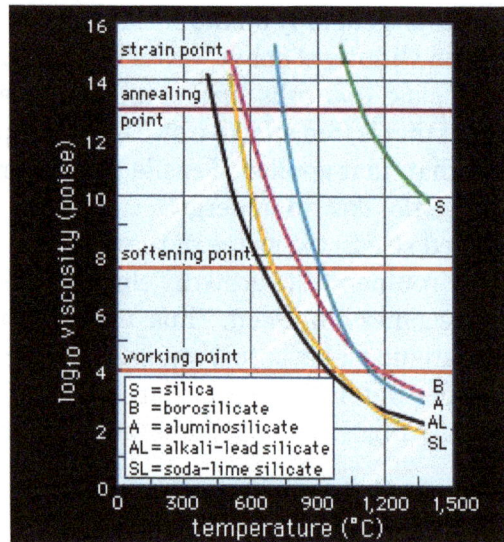

The viscosity of representative silica glasses at varying temperatures.

The annealing point and the strain point lie in the glass transformation range shown in figure; often, the glass transition temperature (T_g) and the annealing point are used synonymously, and the strain point marks the low-temperature end of the range. The T_g may also be considered the maximum temperature for intermittent service. It is evident from figure that the T_g of vitreous silica is the highest of the commercial glasses and that increasing the amount of alkali additions (and therefore the concentration of NWM ions) lowers T_g. Of all the various factors affecting viscosity, water, in the form of hydroxyl ions or molecular water, lowers viscosity the most.

Thermal Expansion

As is evident from figure, glass normally expands when heated and shrinks when cooled. This thermal expansion of glass is critical to its thermal shock performance (that is, its performance when subjected suddenly to a temperature change). When a hot specimen of glass is suddenly cooled—for example, by plunging it in iced water—great tension may develop in the outside layers owing to their shrinking relative to the inner layers. This tension may lead to cracking. Resistance to such thermal shock is known as the thermal endurance of a glass; it is inversely related to the thermal-expansion coefficient and the thickness of the piece.

Soda-lime-silicates and alkali-lead-silicates, which typically have high expansion co-efficients, are quite susceptible to shocking. Improved thermal shock resistance is obtained by using Pyrex-type sodium borosilicates or vitreous silica. For space-based telescopes, the mirror substrates often require materials with expansion coefficients close to zero, in order to avoid any dimensional changes due to temperature fluctuations. A silica glass containing 7.5 percent titanium oxide has a near-zero thermal expansion coefficient and provides satisfactory service in this application.

It should also be evident from figure that the contraction curve of a glass is significantly different from its expansion curve. When glass is used to seal to other materials such as a metal, the relevant parameter is its effective thermal contraction, not its thermal expansion.

Heat Transfer

The thermal conductivity of oxide glass due to atomic vibrations (the so-called pho-non mechanism) does not increase appreciably with temperature. On the other hand, the radiation conductivity (thermal conductivity due to photon transport) increases greatly with temperature. Radiation conductivity is also inversely proportional to the absorption coefficient of a glass for specific photon wavelengths. Thus, the rather high radiation conductivity of molten clear glass enables melting to depths of almost two metres, or five feet, in continuous glass tanks without a serious risk of frozen glass at the bottom. Coloured glasses, on the other hand, have a high photon absorption coefficient and therefore need to be melted either to shallow depths or with electric boosting from the bottom of the tank.

Chemical Properties

The primary determinant of chemical durability in glass is an ion exchange reaction in which alkali ions in the glass are exchanged with hydrogen atoms or hydronium ions present in atmospheric humidity or water. The alkali ions thus leached out of the glass further react with carbon dioxide and water in the atmosphere to produce alkali carbonates and bicarbonates. These are seen as the white deposits that form on a glassy surface in dishwashing tests or after extended humidity exposure (often called weathering). The weathering resistance of several commercial glasses is shown in figure. In general, glasses that are low in alkali offer increased weathering resistance. Vitreous silica is the most resistant, but borosilicates and aluminosilicates also offer excellent weathering resistance.

The leaching mechanism generally operates when the attacking fluid is water or an acidic solution. On the other hand, a dissolution of the entire network may occur when silicate glasses are attacked by caustic alkalis and by hydrofluoric, phosphoric, and perchloric acids. The general approach to improving the chemical durability of glass is to make the surface as silica-rich as possible. This can be accomplished by two methods: fire polishing, a procedure that removes alkali ions by volatilization; or surface treatment with a mixture of sulfur dioxide and steam, which extracts alkali by leaching and converting to washable alkali sulfate. Other methods of improving chemical durability involve limiting the access of water or humidity to the glass surface. Polymeric barrier coatings are effective in this way.

Small amounts of alumina in the glass composition (on the order of 2 to 3 percent) work well to improve the chemical durability of containers. Some high aluminosilicates resist even hot sodium-metal vapours.

Electrical Properties

Electrical Conductivity

Although most glasses contain charged metallic ions capable of carrying an electric current, the high viscosity of glass impedes their movements and electrical activity. Thus, glass is an efficient electrical insulator—though this property varies with viscosity, which in turn is a function of temperature. Indeed, the electrical conductivity of glass increases rapidly with temperature. Hence, in glassmaking it is possible to melt soda-lime-silica glass electrically once it has been heated to about 1,000 °C (1,800 °F) through auxiliary means.

Since univalent alkali ions have the greatest mobility through the glassy structure, they are the primary charge carriers of a glass and therefore determine its electrical conductivity. In general, the higher the concentration of alkalis, the higher the electrical conductivity. The most noted exception from the additivity relationship here is the mixed-alkali effect, in which glasses containing two or more different types of alkali ions have a significantly lower electrical conductivity than linear additivity would suggest. In applications such as high-wattage lamps, where low electrical conductivity is desired, mixed-alkali glasses are useful.

Dielectric Constant

The dielectric, or nonconducting, property of glass is important for its use either as a medium separating the plates of a capacitor or as a substrate in integrated circuits. For capacitor usage, the dielectric constant must be high, whereas for substrates it must be low enough to allow high signal speeds between semiconductor chips. In general, the dielectric constant of glass generally increases with the concentration of NWM ions. Therefore, vitreous silica has one of the lowest dielectric constants, while most soda-lime-silicates have high dielectric constants.

Electronic Conduction

Electronic conduction of charge is important in only two families of glasses: oxide glasses containing large amounts of transition-metal ions and chalcogenides. In metallic solids there are a large number of weakly bound electrons that can move about freely through the crystal structure, but in insulating solids the electrons are confined to specific energy levels known as valence and conduction bands. As the temperature is raised, some electrons from the valence band are able to jump across to the conduction band, thus contributing to what is known as the intrinsic conductivity of the atom. In extrinsic semiconductivity, on the other hand, electrons are provided by defects in the chemical bonding and by impurity atoms. In oxide glasses containing transition-metal ions, for instance, it is believed that electronic conductivity occurs as the hopping of an electron between two transition-metal ions of differing valence that are separated by an oxygen atom. In chalcogenide glasses, semiconductivity is primarily caused by defective bonds in which a particular atom does not follow its covalent coordination.

Optical Properties

Transparency, Opacity and Colour

Because electrons in glass molecules are confined to particular energy levels, they cannot absorb and reemit photons (the basic units of light energy) by skipping from one energy band to another and back again. As a consequence, light energy travels through glass instead of being absorbed and reflected, so that glass is transparent. Furthermore, the molecular units in glass are so small in comparison to light waves of ordinary wavelengths that their absorption effect is negligible. Radiation of some wavelengths, however, can cause glass molecules to vibrate, making the glass opaque to those wavelengths. For instance, most oxide glasses are good absorbers of, and are therefore opaque to, ultraviolet radiation of wavelengths smaller than 350 nanometres, or 3,500 angstroms. These glasses can be made more transparent to ultraviolet radiation by increasing the silica content. At the same time, silicate glasses absorb wavelengths greater than 4 micrometres, making them virtually opaque to infrared radiation. Heavy-metal fluoride glasses, on the other hand, transmit wavelengths up to about 7 micrometres, while some chalcogenide glasses transmit as far as 18 micrometres—properties that make them transparent into the middle infrared region.

Glass to which certain metallic oxides have been added will absorb wavelengths corresponding to certain colours and let others pass, thus appearing tinted to the eye. For instance, cobalt gives an intense blue tint to glass, chromium generally gives green, and manganese imparts purple.

Photosensitivity

In some glasses containing small amounts of cerium oxide and ions of copper, silver, or gold, exposure to ultraviolet radiation causes the oxidation of cerium and the reduction of the latter elements to the metallic state. Upon subsequent heating, the metal nuclei grow, or "strike," developing strong colours (red for copper and gold, yellow for silver) in the ultraviolet-exposed regions of the glass. This technique has been used to produce "three-dimensional photographs," but a more recent use is in microphotolithography for the production of complex electronic circuits.

Traditional photochromic eyeglasses are generally alkali boroaluminosilicates with 0.01 to 0.1 percent silver halide and a small amount of copper. Upon absorption of light, the silver ion reduces to metallic silver, which nucleates to form colloids about 120 angstroms in size. This is small enough to keep the glass transparent, but the colloids are dense enough to make the glass look gray or brown. In photochromic eyeglasses, darkening is reversed either by the removal of light (optical bleaching) or by raising the temperature (thermal bleaching).

Refraction and Reflection of Light

A ray of light, on passing from one transparent medium to another transparent medium of different density, will be transmitted through the second medium with no loss of intensity or change in direction if it strikes the boundary between the two mediums at a right angle (90°). In geometric terms, the right angle at which the light ray meets the boundary is called the normal. If the light ray meets the boundary at an angle other than the normal, then it will be partially reflected back into the first medium and partially refracted, or deflected, in its path through the second medium. The extent to which the light is reflected and refracted depends on the relative densities of the two mediums involved (usually glass and air) and on the angle at which the light ray meets the boundary (known as the angle of incidence). As is shown in figure, if the light ray meets the boundary at less than a certain critical angle (θ_c), most of the light will be refracted while a small amount is reflected. If it arrives at the boundary at the critical angle, then the emerging light will be of diminished intensity and will assume a direction parallel and close to the boundary; most of the light will be reflected. Finally, if the critical angle is exceeded, all of the light will be reflected back into the glass without suffering any loss of intensity. Known as total internal reflection, this phenomenon is widely exploited in single-lens reflex cameras and in fibre optics, in which light signals are piped for great distances before signal boosting is required.

Refraction can be expressed as a constant, known as the refractive index, which is derived mathematically from the ratio of the sine of the angle of incidence on the medium to the sine of the angle of refraction within the medium. The refractive index of a particular type of glass depends on its composition and on the wavelength of the light.

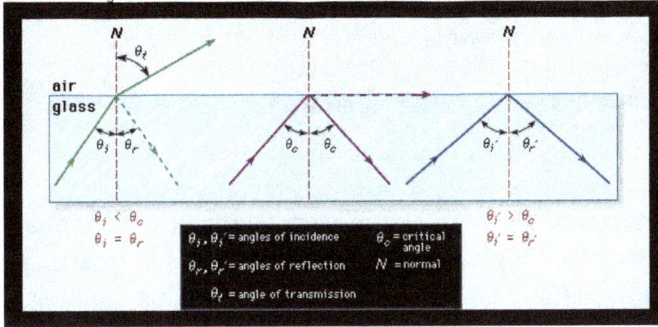

The refraction and reflection of light. (Left) When light strikes the boundary between glass and air at less than the critical angle (θ_c), it is refracted and partially reflected; (centre) when it meets the boundary at the critical angle, it is refracted parallel to the boundary; (right) when it meets the boundary at more than the critical angle, it is reflected totally.

When glass is subjected to unequal stress components operating on perpendicular planes, it becomes birefringent (that is, doubly refracting). The resulting birefringence of a plane-polarized light can be measured by birefringence compensators such as a quartz wedge, and from this measurement the magnitude of the stresses can be estimated. In a polariscope fitted with a tint plate, stressed glass displays colours; the distribution of these colours also may be used for recognizing stress patterns during quality-control operations.

There are various types of glasses used in various purposes of construction and engineering. A few types of such glasses are safety glass, fused quartz, hebron glass, soda lime glass, plate glass, tiffany glass, stained glass, etc. This chapter has been carefully written to provide an easy understanding of these varied types of glass.

SAFETY GLASS

Safety glass is a type of glass that, when struck, bulges or breaks into tiny, relatively harmless fragments rather than shattering into large, jagged pieces. Safety glass may be made in either of two ways. It may be constructed by laminating two sheets of ordinary glass together, with a thin interlayer of plastic, or it may be produced by strengthening glass sheets by heat treatment.

In 1909 the first successful patent for safety glass was taken out in France by an artist and chemist, Édouard Bénédictus, who used a sheet of celluloid bonded between two pieces of glass. Other plastics were also tried, but in 1936 polyvinyl butyral (PVB) was found to possess so many safety-desirable properties that its use became universal. Bulletproof glass is usually built up using several glass and plastic components.

In the heat-treatment method, glass sheets are tempered at about 650 °C (1200 °F), followed by sudden chilling. This treatment increases the strength of the glass sheets approximately sixfold. When such glass does break, it shatters into blunt granules.

SMART GLASS

Smart glass or switchable glass (also smart windows or switchable windows in those applications) is a glass or glazing whose light transmission properties are altered when voltage, light, or heat is applied. In general, the glass changes from transparent to translucent and vice versa, changing from letting light pass through to blocking some (or all) wavelengths of light and vice versa.

Smart glass technologies include electrochromic, photochromic, thermochromic, suspended-particle, micro-blind, and polymer-dispersed liquid-crystal devices.

When installed in the envelope of buildings, smart glass creates climate adaptive building shells.

Electrically Switchable Smart Glass

Suspended-particle Devices

In suspended-particle devices (SPDs), a thin film laminate of rod-like nano-scale particles is suspended in a liquid and placed between two pieces of glass or plastic, or attached to one layer. When no voltage is applied, the suspended particles are randomly organized, thus blocking and absorbing light. When voltage is applied, the suspended particles align and let light pass. Varying the voltage of the film varies the orientation of the suspended particles, thereby regulating the tint of the glazing and the amount of light transmitted.

SPDs can be manually or automatically "tuned" to precisely control the amount of light, glare and heat passing through.

Electrochromic Devices

Electrochromic devices change light transmission properties in response to voltage and thus allow control over the amount of light and heat passing through. In electrochromic windows, the electrochromic material changes its opacity. A burst of electricity is required for changing its opacity, but once the change has been effected, no electricity is needed for maintaining the particular shade which has been reached.

First generation electrochromic technologies tend to have a yellow cast in their clear states and blue hues in their tinted states. Darkening occurs from the edges, moving inward, and is a slow process, ranging from many seconds to several minutes (20–30 minutes) depending on window size. Newer electrochromic technologies eliminate the yellow cast in the clear state and tinting to more neutral shades of gray, tinting evenly rather than from the outside in, and accelerate the tinting speeds to less than three minutes, regardless of the size of the glass. Electrochromic glass provides visibility even in the darkened state and thus preserves visible contact with the outside environment.

Recent advances in electrochromic materials pertaining to transition-metal hydride electrochromics have led to the development of reflective hydrides, which become reflective rather than absorbing, and thus switch states between transparent and mirror-like.

Recent advancements in modified porous nano-crystalline films have enabled the creation of electrochromic display. The single substrate display structure consists of several stacked porous layers printed on top of each other on a substrate modified with a transparent conductor (such as ITO or PEDOT:PSS). Each printed layer has a specific set of functions. A working electrode consists of a positive porous semiconductor such as Titanium Dioxide, with adsorbed chromogens. These chromogens change color by reduction or oxidation. A passivator is used as the negative of the image to improve electrical performance. The insulator layer serves the purpose of increasing the contrast ratio and separating the working electrode electrically from the counter electrode. The counter electrode provides a high capacitance to counterbalances the charge inserted/extracted on the SEG electrode (and maintain overall device charge neutrality). Carbon is an example of charge reservoir film. A conducting carbon layer is typically used as the conductive back contact for the counter electrode. In the last printing step, the porous monolith structure is overprinted with a liquid or polymer-gel electrolyte, dried, and then may be incorporated into various encapsulation or enclosures, depending on the application requirements. Displays are very thin, typically 30 micrometer, or about 1/3 of a human hair. The device can be switched on by applying an electrical potential to the transparent conducting substrate relative to the conductive carbon layer. This causes a reduction of viologen molecules (coloration) to occur inside the working electrode. By reversing the applied potential or providing a discharge path, the device bleaches. A unique feature of the electrochromic monolith is the relatively low voltage (around 1 Volt) needed to color or bleach the viologens. This can be explained by the small over- potentials needed to drive the electrochemical reduction of the surface adsorbed viologens/chromogens.

Polymer-dispersed Liquid-crystal Devices

In polymer-dispersed liquid-crystal devices (PDLCs), liquid crystals are dissolved or dispersed into a liquid polymer followed by solidification or curing of the polymer. During the change of the polymer from a liquid to solid, the liquid crystals become incompatible with the solid polymer and form droplets throughout the solid polymer. The curing conditions affect the size of the droplets that in turn affect the final operating properties of the "smart window". Typically, the liquid mix of polymer and liquid crystals is placed between two layers of glass or plastic that include a thin layer of a transparent, conductive material followed by curing of the polymer, thereby forming the basic sandwich structure of the smart window. This structure is in effect a capacitor.

Electrodes from a power supply are attached to the transparent electrodes. With no applied voltage, the liquid crystals are randomly arranged in the droplets, resulting in scattering of light as it passes through the smart window assembly. This results in the translucent, "milky white" appearance. When a voltage is applied to the electrodes, the electric field formed between the two transparent electrodes on the glass causes the liquid crystals to align, allowing light to pass through the droplets with very little scattering and resulting in a transparent state. The degree of transparency can be controlled by the applied voltage. This is possible because at lower voltages, only a few of the liquid crystals align completely in the electric field, so only a small portion of the light passes through while most of the light is scattered. As the voltage is increased, fewer liquid crystals remain out of alignment, resulting in less light being scattered. It is also possible to control the amount of light and heat passing through, when tints and special inner layers are used.

Micro-blinds

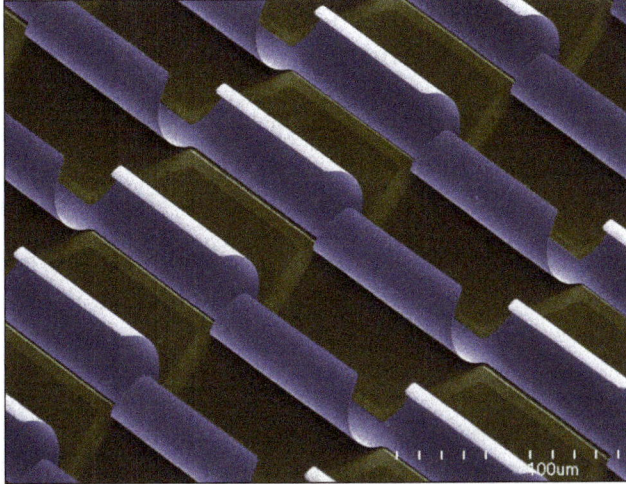

Scanning Electron Microscope (SEM) image of Micro-blinds.

Micro-blinds control the amount of light passing through in response to applied voltage. The micro-blinds are composed of rolled thin metal blinds on glass. They are very small and thus practically invisible to the eye. The metal layer is deposited by magnetron sputtering and patterned by laser or lithography process. The glass substrate includes a thin layer of a transparent conducting oxide (TCO) layer. A thin insulator is deposited between the rolled metal layer and the TCO layer for electrical disconnection. With no applied voltage, the micro-blinds are rolled and let light pass through. When there is a potential difference between the rolled metal layer and the transparent conductive layer, the electric field formed between the two electrodes causes the rolled micro-blinds to stretch out and thus block light. The micro-blinds have several advantages including switching speed (milliseconds), UV durability, customized appearance and transmission.

Related Areas of Technology

The expression smart glass can be interpreted in a wider sense to include also glazings that change light transmission properties in response to an environmental signal such as light or temperature.

- Different types of glazing can show a variety of chromic phenomena, that is, based on photochemical effects the glazing changes its light transmission properties in response to an environmental signal such as light (photochromism), temperature (thermochromism), or voltage (electrochromism).

- Liquid crystals, when they are in a thermotropic state, can change light transmission properties in response to temperature.

- Various metals have been investigated. Thin Mg-Ni films have low visible transmittance and are reflective. When they are exposed to H_2 gas or reduced by an alkaline electrolyte, they become transparent. This transition is attributed to the formation of magnesium nickel hydride, Mg_2NiH_4. Films were created by cosputtering from separate targets of Ni and Mg to facilitate variations in composition. Single-target d.c. magnetron sputtering could be used eventually which would be relatively simple compared to deposition of electrochromic oxides, making them more affordable. The Lawrence Berkeley National Laboratory determined that new transition metals were cheaper and less reactive, but contained the same qualities, thus further reducing the cost.

- Tungsten-doped Vanadium dioxide VO_2 coating reflects infrared light when the temperature rises over 29 °C (84 °F), to block out sunlight transmission through windows at high ambient temperatures.

These types of glazings cannot be controlled manually. In contrast, all electrically switched smart windows can be made to automatically adapt their light transmission properties in response to temperature or brightness by integration with a thermometer or photosensor, respectively.

Examples of Use

Eureka Tower in Melbourne has a glass cube which projects 3 m (10 ft) out from the building with visitors inside, suspended almost 300 m (984 ft) above the ground. When one enters, the glass is opaque as the cube moves out over the edge of the building. Once fully extended over the edge, the glass becomes clear.

The Boeing 787 Dreamliner features electrochromic windows which replaced the pull down window shades on existing aircraft.

NASA is looking into using electrochromics to manage the thermal environment experienced by the newly developed Orion and Altair space vehicles.

Smart glass has been used in some small-production cars including the Ferrari 575 M Superamerica.

ICE 3 high speed trains use electrochromatic glass panels between the passenger compartment and the driver's cabin.

The elevators in the Washington Monument use smart glass in order for passengers to view the commemorative stones inside the monument.

The city's restroom in Amsterdam's Museumplein square features smart glass for ease of determining the occupancy status of an empty stall when the door is shut, and then for privacy when occupied.

ICE 3 high speed train with view into driver's cab.

Same train with glass panel switched to "frosted" mode.

Bombardier Transportation has intelligent on-blur windows in the Bombardier Innovia APM 100 operating on Singapore's Bukit Panjang LRT Line, to prevent passengers from peering into apartments while the train is moving and is planning to offer windows using smart glass technology in its Flexity 2 light rail vehicles.

FUSED QUARTZ

Fused quartz or fused silica is glass consisting of silica in amorphous (non-crystalline) form. It differs from traditional glasses in containing no other ingredients, which are typically added to glass to lower the melt temperature. Fused silica, therefore, has high working and melting temperatures. Although the terms fused quartz and fused silica are used interchangeably, the optical and thermal properties of fused silica are superior to those of fused quartz and other types of glass due to its purity. For these reasons, it finds use in situations such as semiconductor fabrication and laboratory equipment. It transmits ultraviolet better than other glasses, so is used to make lenses and optics for the ultraviolet spectrum. The low coefficient of thermal expansion of fused quartz makes it a useful material for precision mirror substrates.

Manufacture

Fused quartz is produced by fusing (melting) high-purity silica sand, which consists of quartz crystals. There are four basic types of commercial silica glass:

- Type I is produced by induction melting natural quartz in a vacuum or an inert atmosphere.
- Type II is produced by fusing quartz crystal powder in a high-temperature flame.
- Type III is produced by burning $SiCl_4$ in a hydrogen-oxygen flame.
- Type IV is produced by burning $SiCl_4$ in a water vapor-free plasma flame.

Quartz contains only silicon and oxygen, although commercial quartz glass often contains impurities. The most dominant impurities are aluminium and titanium.

Fusion

Melting is effected at approximately 1650 °C (3000 °F) using either an electrically heated furnace (electrically fused) or a gas/oxygen-fuelled furnace (flame-fused). Fused silica can be made from almost any silicon-rich chemical precursor, usually using a continuous process which involves flame oxidation of volatile silicon compounds to silicon dioxide, and thermal fusion of the resulting dust (although alternative processes are used). This results in a transparent glass with an ultra-high purity and improved optical transmission in the deep ultraviolet. One common method involves adding silicon tetrachloride to a hydrogen–oxygen flame, but this precursor results in environmentally unfriendly byproducts including chlorine and hydrochloric acid.

Product Quality

Fused quartz is normally transparent. The material can, however, become translucent if small air bubbles are allowed to be trapped within. The water content (and therefore infrared transmission of fused quartz and fused silica) is determined by the manufacturing process. Flame-fused material always has a higher water content due to the combination of the hydrocarbons and oxygen fuelling the furnace, forming hydroxyl (OH) groups within the material. An IR grade material typically has an (OH) content below 10 ppm.

Applications

Most of the applications of fused silica exploit its wide transparency range, which extends from the UV to the near IR. Fused silica is the key starting material for optical fiber, used for telecommunications.

Because of its strength and high melting point (compared to ordinary glass), fused silica is used as an envelope for halogen lamps and high-intensity discharge lamps, which must operate at a high envelope temperature to achieve their combination of high brightness and long life. Vacuum tubes with silica envelopes allowed for radiation cooling by incandescent anodes.

Because of its strength, fused silica was used in deep diving vessels such as the bathysphere and benthoscope. Fused silica is also used to form the windows of manned spacecraft, including the Space Shuttle and International Space Station.

The combination of strength, thermal stability, and UV transparency makes it an excellent substrate for projection masks for photolithography.

An EPROM with fused quartz window in the top of the package.

Its UV transparency also finds uses in the semiconductor industry; an EPROM, or erasable programmable read only memory, is a type of memory chip that retains its data when its power supply is switched off, but which can be erased by exposure to strong ultraviolet light. EPROMs are recognizable by the transparent fused quartz window which sits on top of the package, through which the silicon chip is visible, and which permits exposure to UV light during erasing.

Due to the thermal stability and composition, it is used in semiconductor fabrication furnaces.

Fused quartz has nearly ideal properties for fabricating first surface mirrors such as those used in telescopes. The material behaves in a predictable way and allows the optical fabricator to put a very smooth polish onto the surface and produce the desired figure with fewer testing iterations. In some instances, a high-purity UV grade of fused quartz has been used to make several of the individual uncoated lens elements of special-purpose lenses including the Zeiss 105 mm f/4.3 UV Sonnar, a lens formerly made for the Hasselblad camera, and the Nikon UV-Nikkor 105 mm f/4.5 (presently soldas the Nikon PF10545MF-UV) lens. These lenses are used for UV photography, as the quartz glass has a lower extinction rate than lenses made with more common flint or crown glass formulas.

Fused quartz can be metallised and etched for use as a substrate for high-precision microwave circuits, the thermal stability making it a good choice for narrowband filters and similar demanding applications. The lower dielectric constant than alumina allows higher impedance tracks or thinner substrates.

Fused quartz is also the material used for modern glass instruments such as the glass harp and the verrophone, and is also used for new builds of the historical glass harmonica. Here, the superior strength and structure of fused quartz gives it a greater dynamic range and a clearer sound than the historically used lead crystal.

Refractory Material Applications

Fused silica as an industrial raw material is used to make various refractory shapes such as crucibles, trays, shrouds, and rollers for many high-temperature thermal processes including steelmaking, investment casting, and glass manufacture. Refractory shapes made from fused silica have excellent thermal shock resistance and are chemically inert to most elements and compounds, including virtually all acids, regardless of concentration, except hydrofluoric acid, which is very reactive even in fairly low concentrations. Translucent fused-silica tubes are commonly used to sheathe electric elements in room heaters, industrial furnaces, and other similar applications.

Owing to its low mechanical damping at ordinary temperatures, it is used for high-Q resonators, in particular, for wine-glass resonator of hemispherical resonator gyro.

Quartz glassware is occasionally used in chemistry laboratories when standard borosilicate glass cannot withstand high temperatures or when high UV transmission is required. The cost of production is significantly higher, limiting its use; it is usually found as a single basic element, such as a tube in a furnace, or as a flask, the elements in direct exposure to the heat.

Physical Properties

The extremely low coefficient of thermal expansion, about $5.5 \cdot 10^{-7}/K$ (20...320 °C), accounts for its remarkable ability to undergo large, rapid temperature changes without cracking.

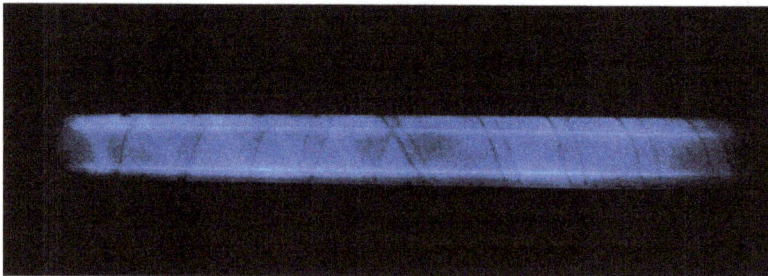

Phosphorescence in fused quartz from an extremely intense pulse
of UV light in a flashtube, centered at 170 nm.

Fused quartz is prone to phosphorescence and "solarisation" (purplish discoloration) under intense UV illumination, as is often seen in flashtubes. "UV grade" synthetic fused silica (sold under various tradenames including "HPFS", "Spectrosil", and "Suprasil") has a very low metallic impurity content making it transparent deeper into the ultraviolet. An optic with a thickness of 1 cm has a transmittance around 50% at a wavelength of 170 nm, which drops to only a few percent at 160 nm. However, its infrared transmission is limited by strong water absorptions at 2.2 μm and 2.7 μm.

"Infrared grade" fused quartz (tradenames "Infrasil", "Vitreosil IR", and others), which is electrically fused, has a greater presence of metallic impurities, limiting its UV transmittance wavelength to around 250 nm, but a much lower water content, leading to excellent infrared transmission up to 3.6 μm wavelength. All grades of transparent fused quartz/fused silica have nearly identical physical properties.

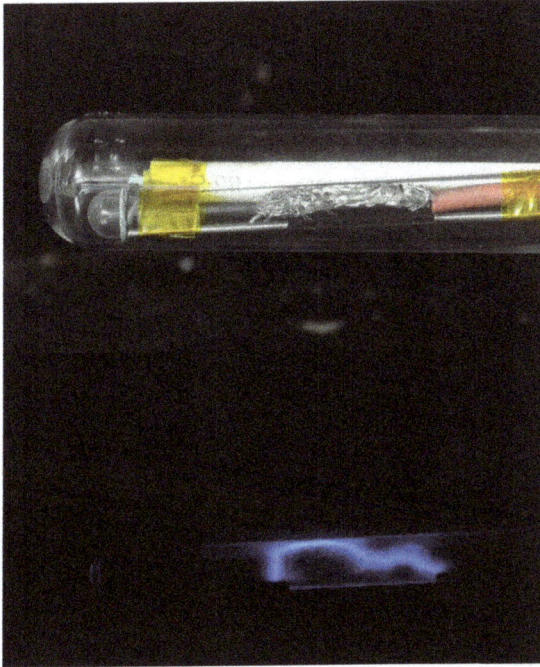

Phosphorescence of the quartz ignition tube of an air-gap flash.

Optical Properties

The optical dispersion of fused silica can be approximated by the following Sellmeier equation:

$$\varepsilon = n^2 = 1 + \frac{0.6961663\lambda^2}{\lambda^2 - 0.0684043^2} + \frac{0.4079426\lambda^2}{\lambda^2 - 0.1162414^2} + \frac{0.8974794\lambda^2}{\lambda^2 - 9.896161^2}$$

where the wavelength λ λ is measured in micrometers. This equation is valid between 0.21 and 3.71 μm and at 20 °C. Its validity was confirmed for wavelengths up to 6.7 μm.

Typical Properties of Clear Fused Silica

- Density: 2.203 g/cm³.

- Hardness: 5.3 to 6.5 (Mohs scale), 8.8 GPa.

- Tensile strength: 48.3 MPa.

- Compressive strength: >1.1 GPa.

- Bulk modulus: ~37 GPa.

- Rigidity modulus: 31 GPa.

- Young's modulus: 71.7 GPa.

- Poisson's ratio: 0.17.

- Lamé elastic constants: λ = 15.87 GPa, μ = 31.26 GPa.

- Coefficient of thermal expansion: $5.5 \cdot 10^{-7}$/K (average from 20 to 320 °C).

- Thermal conductivity: 1.3 W/(m·K).

- Specific heat capacity: 45.3 J/(mol·K).

- Softening point: $\approx$1665 °C.

- Annealing point: $\approx$1140 °C.

- Strain point: 1070 °C.

- Electrical resistivity: $>10^{18}$ Ω·m.

- Dielectric constant: 3.75 at 20 °C 1 MHz.

- Magnetic susceptibility: $-11.28 \cdot 10^{-6}$ (SI, 22 °C).

- Dielectric loss factor: less than 0.0004 at 20 °C 1 MHz typically $6 \cdot 10^{-5}$ at 10 GHz.

- Index of refraction: n_d = 1.4585 (at 587.6 nm).

- Change of refractive index with temperature (0 to 700°C): $1.28 \cdot 10^{-5}$/K (between 20 to 30 °C).

- Strain-optic coefficients: p_{11} = 0.113, p_{12} = 0.252.

- Hamaker constant: A = $6.5 \cdot 10^{-20}$ J.

- Dielectric strength: 250 to 400 kV/cm at 20 °C.

- Surface tension: 0.300 N/m at 1800 to 2400 °C.

- Abbe Number: Vd = 67.82.

PHOTOSENSITIVE GLASS

Photosensitive glass, also known as photostructurable glass (PSG), or photomachinable glass, is a crystal-clear glass that belongs to the lithium-silicate family of glasses, in which an image of a mask can be captured by microscopic metallic particles in the glass when it is exposed to short wave radiations such as ultraviolet light. Photosensitive glass was first discovered by S. Donald Stookey in 1937.

Exposure Process

When the glass is exposed to UV light in the wavelength range 280–320 nm, a latent image is formed. The glass remains transparent at this stage, but its absorption in the uv range of the spectrum increases. This increased absorption is only detectable using uv transmission spectroscopy. The reason behind this is suggested to be an oxidation reduction reaction that occurs inside the glass during exposure in which cerium ions are oxidized to a more stable state and silver ions are reduced to silver.

Post-exposure Heat Treatment

When the glass is heated to temperatures in the range 550–560 °C for several hours the latent image is converted to a visible image through photoexcitation. Exposure through photographic negatives permits the development of three-dimensional color images and photographs. This heat treatment is done in two stages: the temperature is first raised to about 500 °C to allow for the completion of the oxidation-reduction reaction, and formation of silver nanoclusters. In the following stage, when the temperature is raised to 550–560 °C, a new material (lithium metasilicate) with the formula (Li_2SiO_3) forms on the silver nanoclustors, this material forms in the crystalline phase.

HF Chemical Etching

The lithium metasilicate that forms in the exposed regions of the glass has the unique property of being strongly etched in hydrofluoric acid (HF). Hence allowing a three-dimensional image of the mask to be produced, the resulting glass microstructures have a surface roughness in the range 5 μm to 0.7 μm.

X-ray Sensitive Glasses

As stated above radiation produces some direct and indirect measurable changes in the glasses. In some cases, the effect is readily observable immediately upon irradiation. In other cases, thermal treatment is required to bring about the observed changes. On the whole, the result of the mentioned reactions will be atomic silvers and silver clusters which act as nucleant for precipitation of lithium-meta-silicate during post heat-treatment of irradiated glass and Similar to other glass-ceramic systems, the more nucleation sites leads to more reduction of crystallization temperature and finer crystalline size. Therefore, to attain the above-mentioned condition, various energetic radiation such as UVand laser beam and x γ and proton and radiations have been used for different photosensitive glasses until now. Imanieh et al. investigated the effect of X-ray irradiation on solarization of photosensitive lithium silicate based glasses containing cerium, antimony, tin and silver elements. They have shown that there is a possibility to use X-ray in photosensitive glasses. This will open new doors for nano machining of glasses in near future.

Applications

Photosensitive glass is used in printing and reproducing processes. Photosensitive glass is like traditional camera film except that it reacts to ultraviolet (UV) light, where camera film responds to visible light. The ideal wavelength to use for exposure should be between 300 and 350 nm, with 320 nm being optimum.

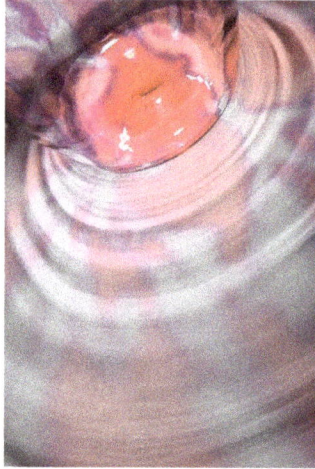

Photosensitive glass.

Spinfex photosensitive glass.

Photosensitive glass contains microscopic metallic particles. These microscopic metallic ion nanoparticles are made of gold or silver which is responsible for the refractive index change. Photosensitive glass is similar to photographic film. Photographic film uses chemicals, while photosensitive glass uses gold or silver ions in the material that will respond to the action of light. The process is to pass light of the ultraviolet wavelength through a negative on the glass. Photographic resolution can be obtained with adhesive polyester as a reverse negative, however anything which resists UV light can act as a "negative."

The glass is sensitive to light that when passes through a mask can ultimately turn it into a permanent picture with a heat process "fixing" permanently the image. Silver glass "latent images" will develop in 3–4 hours at 886–976 °F. Gold glass "latent images" require a higher temperature of 968–1058 °F and over a similar period of time for postbaking. Postbaking hastens the occurrence of the particles with the shadow areas of the negative, permitting deeper penetration into the glass than the highlighted areas. This gives the picture three dimensions and color.

The photograph is developed by heating the photosensitive glass around 1000 °F for several hours after exposure. The glass itself is photosensitive and produces a three-dimensional image. Particles that are invisible to the naked eye (i.e. gold or silver) are in the glass. These microscopic particles move and grow when heated to form the photographic image itself. The process is similar to camera film, however a "negative" is placed on top of the photosensitive glass and then exposed to "'ultraviolet'" light. Camera film, of course, would be exposed to ordinary visible light. Then there is a special process for the exposed photosensitive glass. The glass is reheated then in a kiln and postbaked for several hours. The image then "appears" within the special exposed glass as if by magic. The heated piece of photosensitive glass is then allowed to cool down and the process is done. The positive images produced within photosensitive glass comes in a variety of colors.

As a material for the hot glass studio artist, an additional method of producing imagery in an object using photosensitive glass is to first blow an extremely thin rondel (cased or otherwise) which is annealed in the typical manner. That rondel is then cut into sections which are exposed under a negative. Next, those sections (containing the latent image) are warmed and applied to the surface of a gather of hot glass on the blowpipe. As the object is completed over several furnace reheats, the heat develops the image as the object is being created. This method specifically eliminates the need for the reheating of the object in a kiln for development, which consumes considerable oven time, energy, and the risk of loss or damage due to shattering on the way up to temperature, or more importantly, slumping while being held at temperature. The timing of the glassblower determines the final degree of development, and simple choices of form minimize distortion in the image.

Since the image is inside and actually a part of the glass itself, photosensitive glass is the most durable photographic medium known. It is claimed that a photo image within photosensitive glass is the most durable form of photography and will last as long as the glass itself. The photographic image is not on the surface of the glass, but internally.

Fluorescent photosensitive glass makes it possible to make fluorescent photographs and fluorescence holography.

Photosensitive glass is different from photochromatic glass. Photochromatic glass is used in self-darkening sunglasses which darkens when exposed to bright daylight. It then returns to see-through transparency when strong daylight is removed and can then be used indoors as regular glasses.

Little has been done to develop the product since its patent. It is labor-intensive and has a high cost. Only large commercial glass factories produce it. In the 1980s photosensitive glass was created to a small degree to be used in "hot glass" work. Then individual artists owned smaller studios and created works in blown glass and began experimenting with photosensitive glass. Going into the Twenty-First Century only a few glass artists know the technique of achieving good results with photosensitive glass. In the present time, the only photosensitive glasses produced are PhotoCor(R), Foturan and APEX. PhotoCor is produced by the inventor, Corning, Inc. Foturan is produced by SCHOTT Corporation and APEX by Life Bioscience. Photosensitive glass has been used as a holographic material to record diffractive optical elements for high power laser applications.

Military Applications

One of the reasons for the delay between the invention of photosensitive glass and its public announcement approximately ten years later was its potential use in military applications. It is possible to burn images and words that are hidden in photosensitive glass until heated at a high temperature. The military used this fact during World War II to send secret messages to allied troops in pieces of what looked like "ordinary glass". At the other end, the person who received the "ordinary glass" just had to heat it up to read the hidden message. Because of this application photosensitive glass was kept secret until the end of World War II.

Building Applications

The United Nations Building is faced with hundreds of square feet of photosensitive glass.

FOTURAN

Foturan (notation of the manufacturer: FOTURAN) is a photosensitive glass by SCHOTT Corporation developed in 1984. It is a technical glass-ceramic which can be structured without photoresist when it is exposed to shortwave radiation such as ultraviolet light and subsequently etched.

In February 2016, Schott announced the introduction of Foturan II at Photonics West. Foturan II is characterized by higher homogeneity of the photosensitivity which allows finer microstructures.

Composition and Properties

Composition										
Ingre-dient	SiO_2	LiO_2	Al_2O_3	K_2O	Na_2O	ZnO	B_2O_3	Sb_2O_3	Ag_2O	CeO_2
Share [%]	75-85	7-11	3-6	3-6	1-2	0-2	0-1	0,2-1	0,1-0,3	0,01-0,2

Mechanical Properties	
Knoop-Hardness in N/mm^2 (0.1/20)	480
Vickers-Härte in N/mm^2 (0.2/25)	520
Density in g/cm^3	2,37

Thermal Properties	
Coefficient of mean linear thermal expansion a_{20-300} in $10^{-6} \cdot K^{-1}$	8,49
Thermal Conductivity at 90 °C in W/mK	1,28
Transformation Temperature T_g in °C	455

Electrical Properties

Relative Permittivity

Frequency [GHz]	1.1	1.9	5
Glass-state (annealed at 40 °C/h)	6.4	6.4	6.4
Ceramic-state (ceramized at 560 °C)	5.8	5.9	5.8
Ceramic-state (ceramized at 810 °C)	5.4	5.5	5.4

Dissipation factor $\tan\alpha (\cdot 10^{-4})$

Frequency [GHz]	1.1	1.9	5
Glass-state (annealed at 40 °C/h)	84	90	109
Ceramic-state (ceramized at 560 °C)	58	65	79
Ceramic-state (ceramized at 810 °C)	39	44	55

Chemical Properties

Hydrolytic resistance acc. to DIN ISO 719 in $\mu g Na_2O/g$ (class)	5 7 8 (HGB 4)
Acid resistance acc. to DIN 12116 in mg/dm^2 (class)	0,48 (S1)
Alkali resistance acc. to DIN ISO 695 in mg/dm^2 (class)	100 (A2)

Optical properties

Refractive Index

wavelength [nm], $\lambda=$	300	486.1 (n_F)	546.1 (n_e)	567.6 (n_d)	656.3 (n_C)
Glass-state (annealed at 40 °C/h)	1.549	1.518	1.515	1.512	1.510
Ceramic-state (ceramized at 560 °C)	n/a	1.519	1.515	1.513	1.511
Ceramic-state (ceramized at 810 °C)	n/a	1.532	1.528	1.526	1.523
Spectral Transmittance					
$\tau(\lambda)$	t_{250}	t_{270}	t_{280}	t_{295}	t_{350}
in [%, 1mm]	0.1	3	11	29	89

Foturan is a lithium aluminosilicate glass system doped with small amounts of silver oxides and cerium oxides.

Processing

Foturan can be structured via UV-exposure, tempering and etching: Crystal nucleation grow in Foturan when exposed to UV and heat treated afterwards. The crystalized areas react much faster to hydrofluoric acid than the surrounding vitreous material, resulting in very fine microstructures, tight tolerance and high aspect ratio.

Exposure

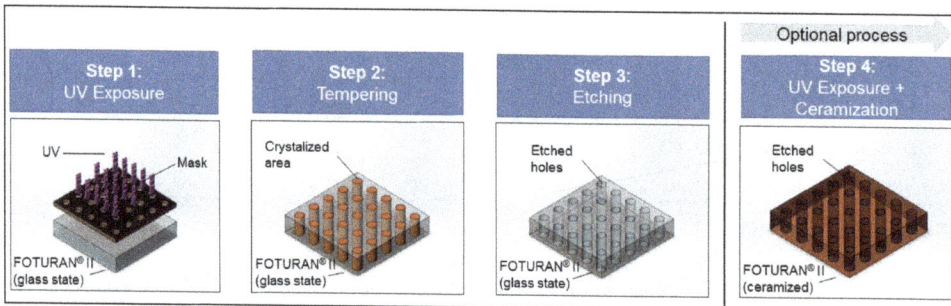

Foturan - Processing steps (schematic view).

If Foturan is exposed to light in the ultra-violet-range with a wavelength of 320 nm (eventually via photomask, contact lithography or proximity lithography to expose certain patterns), a chemical reaction is started in the exposed areas: The containing Ce^{3+} transforms into Ce^{4+} and frees an electron.

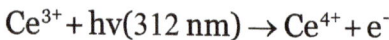

$$Ce^{3+} + h v (312 \, nm) \rightarrow Ce^{4+} + e^-$$

Tempering

During the nucleation tempering ($\sim$ 500 °C), the Silver-ion Ag^+ will be transferred into Ag^0 by scavenging the electron released from Ce^{3+}.

$$Ag^+ + e^- + DH \rightarrow Ag^0$$

This activates the agglomeration of atomic silver to form nanometer-scale silver clusters,

$$xAg^0 + DH \rightarrow (Ag^0)x$$

During the subsequent crystallization tempering ($\sim$560-600 °C), lithium metasilicates (Li_2SiO_3 glass-ceramic) forms on the silver cluster nucleation in the exposed areas. The un-exposed glass, otherwise amorphous, remains unchanged.

Etching

After tempering, the crystallized areas can be etched with hydrofluoric acid 20 times faster than the unexposed, still amorphous glass. Thus, structures with an aspect ratio of ca. 10:1 can be created.

Ceramization

After etching, a ceramization of the entire substrate after a 2nd UV-exposure and thermal treatment is possible. The crystalline phase in this stage is lithium dicilicate $Li_2Si_2O_5$.

Product Characteristics

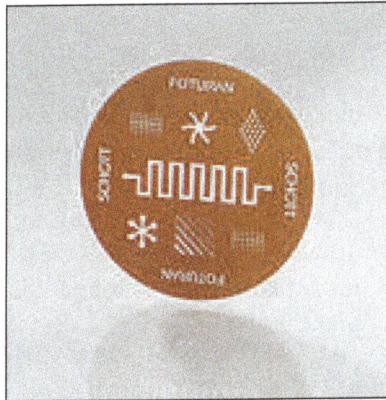

Foturan (ceramized).

- Small structure size: Structure sizes of $\sim$ 25 μm are possible.

- High aspect ratio: Etchingratios of > 20:1 make aspect ratio of > 10:1 and a wall angle of $\sim$ 1-2° possible.

- High optical transmission in visible and non-visible spectrum: More than 90% transmission (substrate thickness 1 mm) between 350 nm and 2.700 nm.

- High temperature resistance: Tg > 450°Celsius.

- Pore-free: Suitable for biotech / microfluidics application.

- Low self fluorescence.

- Hydrolytic resistance (acc. to DIN ISO 719): HGB 4.

- Acid resistance (acc. to DIN 12116): S 1.

- Alkali resistance (acc. to DIN ISO 695): A 2.

Foturan in the Scientific Community

Foturan is a widely known material in the material science community. As of October 30, 2015, Google Scholar showed more than 1.000 results of Foturan in scholarly literatures across an array of publishing formats and disciplines.

Many of those deal with topics such as:

- Micromachining Foturan.

- 3D / laser direct writing in Foturan.

- Using Foturan for optical waveguides.

- Using Foturan for volume gratings.

- Processing Foturan via excimer / femtosecond laser.

Applications

Foturan is mainly used for microstructure applications, where small and complex structures have to be created out of a solid and robust base material. Overall there are five main areas for which Foturan is used:

- Microfluidics / Biotech (such as lab-on-a-chip or organ-on-a-chip components, micro mixer, micro reactor, printheads, titer plates, chip electrophoresis).

- Semiconductor (such FED spacer, packaging elements or interposer for IC components, CMOS or memory modules).

- Sensors (such as flow- or temperature sensors, gyroscopes or accelerometers).

- RF / MEMS (such as substrates or packaging elements for antennas, capacitors, filter, duplexers, switches or oscillators).

- Telecom (such as optical alignment chips, optical waveguides or optical interconnects).

By thermal diffusion bonding it is possible to bond multiple Foturan layers on top of each other to create complex 3-dimensional microstructures.

CER-VIT

Cer-Vit is a family of glass-ceramic materials that were invented by Owens Illinois in the mid-1960s. Its principle ingredients are the oxides of lithium, aluminum and silicon. It is melted to form a glass which is then heat treated to nucleate and crystallized it into a material that is more than 90% microscopic crystals. Its formulation and heat treatment can be modified to produce a variety of material properties. One form is a material that is transparent and has a near zero thermal expansion. Its transparency is because the microscopic crystals are smaller than the wave length of light and are transparent, and its low thermal expansion is because they have a spodumene structure.

This material (Cer-Vit C 101) was used to form large mirror blanks (158 inches in diam-eter) that were used in telescopes in several places, including South America, France and Australia. Owens Illinois ceased production of C101 in 1978. In addition, Cer-Vit materials were used to make stove tops, cook ware and aviation applications, but never commercialized.

Today, glass-ceramic products such as transparent mirror blanks and stove tops, and cookware are manufactured and in daily use. These products include trade names of Zerodor, Hercuvit, and Pyroceram. Most of which have low or zero thermal expansion, which allows them to be exposed to rapid temperature changes or localized heating or cooling.

Applications

At Mount Lemmon Observatory, two 1.5 meter diameter telescopes have a Cer-Vit glass mirror. One of the telescopes discovered 2011 AG5, an asteroid which achieved 1 on the Torino Scale.

POROUS GLASS

Porous glass is glass that includes pores, usually in the nanometre- or micrometre-range, commonly prepared by one of the following processes: through metastable phase separation in borosilicate glasses (such as in their system SiO_2-B_2O_3-Na_2O), followed by liquid extraction of one of the formed phases; through the sol-gel process; or simply by sintering glass powder.

The specific properties and commercial availability of porous glass make it one of the most extensively researched and characterized amorphous solids. Due to the possibility of modeling the microstructure, porous glasses have a high potential as a model system. They show a high chemical, thermal and mechanical resistance, which results from a rigid and incompressible silica network. They can be produced in high quality and with pore sizes ranging from 1 nm up to any desired value. An easy functionalization of the inner surface opens a wide field of applications for porous glasses.

A further special advantage of porous glasses compared to other porous materials, is that they can be made not only as powder or granulate, but also as larger pieces in almost any user defined shape and texture.

In scientific literature, porous glass is a porous material containing approximately 96% silica, which is produced by an acidic extraction or a combined acidic and alkaline extraction respectively, of phase separated alkali borosilicate glasses, and features a three-dimensional interconnected porous microstructure. For commercially available porous glasses, the terms porous VYCOR-Glass (PVG) and Controlled Pore Glass (CPG) are used. The pore structure is formed by a syndetic channel system and has a specific surface from 10 to 300 m²/g. Porous glasses can be generated by an acidic extraction of phase separated alkaliborosilica glasses, or by a sol-gel-process. By regulating the manufacturing parameters, it is possible to produce a porous glass with a pore size of between 0.4 and 1000 nm in a very narrow pore size distribution. You can generate various moulds, for example, irregular particles (powder, granulate), spheres, plates, sticks, fibers, ultra thin membranes, tubes and rings.

Manufacturing

Precondition for repetitious manufacturing of porous glass is the knowledge about structure determining and structure controlling parameters. The composition of the initial glass is a structure controlling parameter. The manufacturing of the initial glass, mainly the cooling process, the temperature and time of thermal treatment, and the after treatment are structure determining parameters. The phase diagram for sodium-borosilica glass shows a miscibility gap for certain glass compositions.

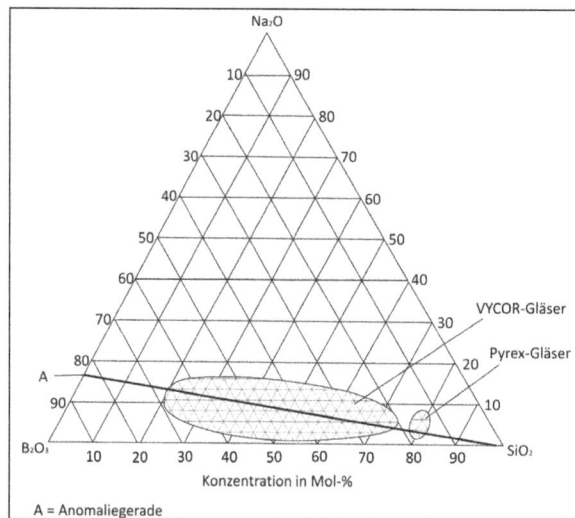

Ternary Phase diagram in the sodium borosilicate system.

The upper critical temperature lies at about 760 °C and the lower one at about 500 °C. O.S. Moltschanova was the first person who exactly described the definition of the exsolution. For a phase separation the initial glass composition must lie in the miscibility gap of the ternary Na_2O-B_2O_3-SiO_2 glass system. By a thermal treatment, an interpenetration structure is generated, which results from a spinodal decomposition of the sodium-rich borate phase and the silica phase. This procedure is called primary decomposition. Using an initial glass composition, which lies on the line of anomaly, it is possible to attain a maximum decomposition, which is almost strainless.

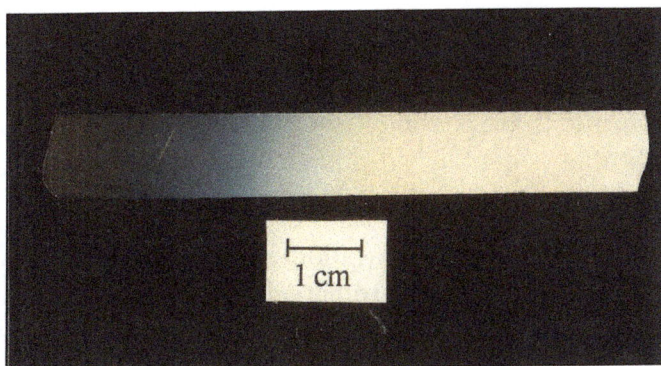

Porous glass filled with water, sample about 1 mm thick, made by phase separation in a thermal gradient (high temperature on the right) of a s odium borosilicate glass, followed by acid leaching.

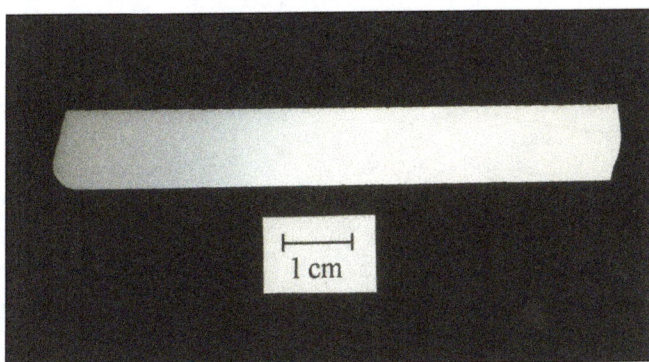

Same porous glass as above, but dry. The increased difference between the refractive indices glass/air as compared to glass/water is causing greater whiteness based on the Tyndall effect.

As both phases have a different resistances to water, mineral acids, and inorganic salt solutions, the sodium-rich borate phase in these mediums can be removed by extraction. Optimal extraction is possible only if the initial glass composition and thermal treatment are chosen such that combine structures form, and not droplet structures. The texture is influenced by the composition of the initial glass, which directs size and type of decomposition areas. In the context of porous glasses, "texture" implies properties like specific pore volume, specific surface, pore size, and porosity. The emerging areas of decomposition depend on time and temperature of the thermal treatment.

Furthermore, the texture of porous glasses is influenced by the concentration of the extraction medium and the ratio of fluid to solid.

Also, colloidal silica is solving in the sodium-rich borate phase, when time and temperature of thermal treatment are increased. This process is called secondary decomposition. The colloidal silica deposit in the macro pores during extraction and obscure the real pore structure. The solubility of colloidal silica in alkaline solutions is higher than network silica, and thus can be removed by an alkaline after-treatment.

Applications

Because of their high mechanical, thermal and chemical stability, variable manufacturing of pore sizes with a small pore size distribution and variety of surface modifications, a wide array of applications are possible. The fact that porous glasses can be produced in many different shapes is another advantage for application in industry, medicine, pharmacy research, biotechnology and sensor technology.

Porous glasses are ideal for material separation, because of the small pore size distribution. This is why they are used in gas chromatography, thin layer chromatography and affinity chromatography. An adaptation of stationary phase for a separation problem is possible by a specific modification of the surface of the porous glass.

In biotechnology, porous glasses have benefits for the cleaning of DNA and the immobilization of enzymes or microorganisms. Controlled pore glass (CPG) with pore sizes between 50 and 300 nm is also excellently suited for the synthesis of oligonucleotides. In this application, a linker, a nucleoside or a non-nucleosidic compound, is first attached to the surface of CPG. The chain length of produced oligonucleotides is dependent on the pore size of CPG.

In addition, porous glasses are used for manufacturing implants, especially dental implants, for which porous glass powder is processed with plastics to form a composite. The particle size and the pore size influence the elasticity of the composite so as to fit the optical and mechanical properties to surrounding tissue, for example, the appearance and hardness of dental enamel.

With the ability to form porous glasses as platelets, membrane technology is another important area of application. Hyper filtration of sea – and brackish water and ultra filtration in "downstream process" are but two. Additionally, they are often appropriate as a carrier for catalysts. For example, the olefin – metathesis was realized on the system metal – metal oxide/porous glass.

Porous glasses can be used as membrane reactors as well, again because of their high mechanical, thermal and chemical stability. Membrane reactors can improve conversion of limited balance reactions, while one reaction product is removed by a selective membrane. For example, in the decomposition of hydrogen sulfide on a catalyst in a glass capillary, the conversion by reaction was higher with glass capillary than without.

TEMPERATURE SENSITIVE GLASS

Temperature sensitive glass is a glass material that reacts to ambient temperatures radiated off of other surfaces, e.g. hands or water. The liquid crystals beneath the glass surface impact color upon temperature. There are three main phases of these crystals: nematic, smectic, and chiral.

Process

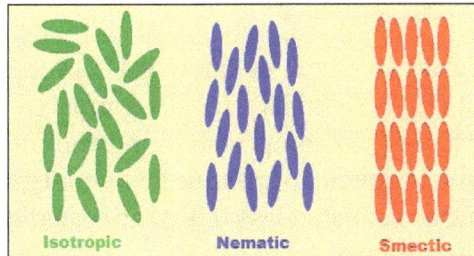

Phases of Liquid Crystals.

Liquid Crystal Phases

Visual glass goes through a gradual progression while altering colors stages in different heat zones. In order for specific light wavelengths to be reflected off of a temperature sensitive glass, it has to go through one of three main heat phases. In accelerated temperature zones, the crystals respond in the nematic phase. Smectic is in the range of temperatures between that of its neighbors nematic and chiral. These phases are impacted by the pitch which in return reflects specific wavelengths.

Pitch Effects

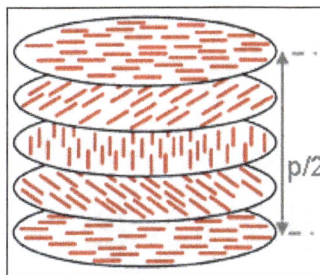

PitchPlanes.

Temperature changes the distance between the pitch planes. Pitch tightens with an increase temperature and expands when temperature plummets.

The crystal's pitch is the distance it takes one crystal to make one complete rotation. This determines the wavelength of light that will be reflected and therefore determines the color. The pitch is equal to the corresponding wavelength of light.

Applications

The physical uses of temperature sensitive glass is mostly in a visual context. Some shower walls use these glass tiles on the walls so users can observe how hot/cold the water is before entry. Other applications include mood rings, battery condition display and coffee cups.

TEMPERED GLASS

Tempered or toughened glass is a type of safety glass processed by controlled thermal or chemical treatments to increase its strength compared with normal glass. Tempering puts the outer surfaces into compression and the interior into tension. Such stresses cause the glass, when broken, to crumble into small granular chunks instead of splintering into jagged shards as plate glass (a.k.a. annealed glass) does. The granular chunks are less likely to cause injury.

As a result of its safety and strength, tempered glass is used in a variety of demanding applications, including passenger vehicle windows, shower doors, architectural glass doors and tables, refrigerator trays, mobile phone screen protectors, as a component of bulletproof glass, for diving masks, and various types of plates and cookware.

Properties

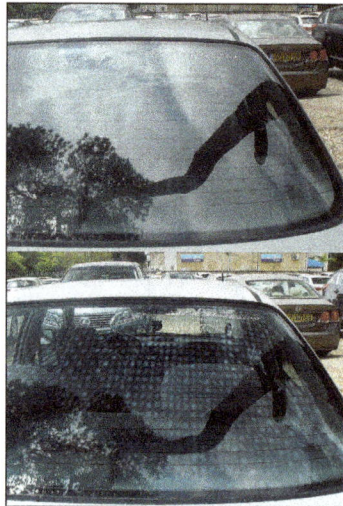

Tempered glass of car rear window. Variations in glass stress are clearly seen when photographed through a polarizing filter.

Tempered glass is physically and thermally stronger than normal glass. The greater contraction of the inner layer during manufacturing induces compressive stresses in the surface of the glass balanced by tensile stresses in the body of the glass. For glass to be considered tempered, this compressive stress on the surface of the glass should be a minimum of 69 megapascals (10,000 psi). For it to be considered safety glass, the surface compressive stress should exceed 100 megapascals (15,000 psi). As a result of the increased surface stress, if the glass is ever broken it only breaks into small circular pieces as opposed to sharp jagged shards. This characteristic makes tempered glass safe for high-pressure and explosion proof applications.

It is this compressive stress that gives the tempered glass increased strength. This is because annealed glass, which has almost no internal stress, usually forms microscopic surface cracks, and in the absence of surface compression, any applied tension to the glass causes tension at the surface, which can drive crack propagation. Once a crack starts propagating, tension is further concentrated at the tip of the crack, causing it to propagate at the speed of sound in the material. Consequently, annealed glass is fragile and breaks into irregular and sharp pieces.

Any cutting or grinding must be done prior to tempering. Cutting, grinding, and sharp impacts after tempering will cause the glass to fracture.

The strain pattern resulting from tempering can be observed by viewing through an optical polarizer, such as a pair of polarizing sunglasses.

Uses

Tempered glass is used when strength, thermal resistance, and safety are important considerations. Passenger vehicles, for example, have all three requirements. Since they are stored outdoors, they are subject to constant heating and cooling as well as dramatic temperature changes throughout the year. Moreover, they must withstand small impacts from road debris such as stones as well as automobile accidents. Because large, sharp glass shards would present additional and unacceptable danger to passengers, tempered glass is used so that if broken, the pieces are blunt and mostly harmless. The windscreen or windshield is instead made of laminated glass, which will not shatter into pieces when broken while side windows and the rear windshield are typically tempered glass.

Safety approval markings on an automobile vent
window made for a Chrysler car by PPG.

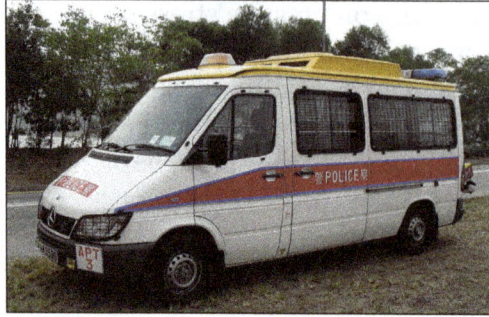
Police van with screen protector.

Other typical applications of tempered glass include:

- Balcony doors,

- Athletic facilities,

- Swimming pools,

- Facades,

- Shower doors and bathroom areas,

- Exhibition areas and displays,

- Computer towers or cases.

Buildings and Structures

Tempered glass is also used in buildings for unframed assemblies (such as frameless glass doors), structurally loaded applications, and any other application that would become dangerous in the event of human impact. Tempered and heat strengthened glass can be three to seven times stronger than annealed glass. Building codes in the United States require tempered or laminated glass in several situations including some skylights, near doorways and stairways, large windows, windows which extend close to floor level, sliding doors, elevators, fire department access panels, and near swimming pools.

Household Uses

Tempered glass is also used in the home. Some common household furniture and appliances that use tempered glass are frameless shower doors, glass table tops, replacement glass, glass shelves, cabinet glass and glass for fireplaces.

Food Service

"Rim-tempered" indicates that a limited area, such as the rim of the glass or plate, is tempered and is popular in food service. However, there are also specialist manufacturers that offer a fully tempered/toughened drinkware solution that can bring increased benefits in the form of strength and thermal shock resistance. In some countries these products are specified in venues that require increased performance levels or have a requirement for a safer glass due to intense usage.

Tempered glass has also seen increased usage in bars and pubs, particularly in the United Kingdom and Australia, to prevent broken glass being used as a weapon. Tempered glass products can be found in hotels, bars, and restaurants to reduce breakages and increase safety standards.

Cooking and Baking

Some forms of tempered glass are used for cooking and baking. Manufacturers and brands include Glasslock, Pyrex, Corelle, and Arc International. This is also the type of glass used for oven doors.

Mobile Devices

Most touchscreen mobile devices use some form of toughened glass (such as Corning's Gorilla Glass), as do some aftermarket screen protectors for these devices.

Manufacturing

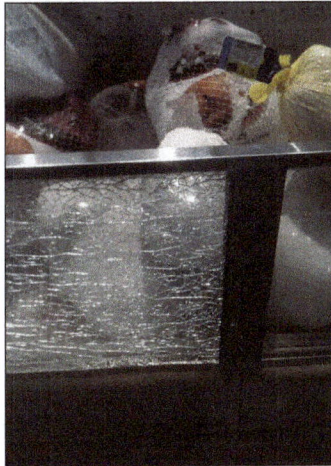

Tempered safety glass which has been laminated often does not fall
out of its frame when it breaks – usually due to the anti-splinter film
applied on the glass, as seen in this grocery store meat case.

Tempered glass can be made from annealed glass via a thermal tempering process. The glass is placed onto a roller table, taking it through a furnace that heats it well above its transition temperature of 564 °C (1,047 °F) to around 620 °C (1,148 °F). The glass is then rapidly cooled with forced air drafts while the inner portion remains free to flow for a short time.

An alternative chemical toughening process involves forcing a surface layer of glass at least 0.1 mm thick into compression by ion exchange of the sodium ions in the glass surface with potassium ions (which are 30% larger), by immersion of the glass into a bath of molten potassium nitrate. Chemical toughening results in increased toughness compared with thermal tempering and can be applied to glass objects of complex shapes.

Disadvantages

Tempered glass must be cut to size or pressed to shape before tempering, and cannot be re-worked once tempered. Polishing the edges or drilling holes in the glass is carried out before the tempering process starts. Because of the balanced stresses in the glass, damage to any portion will eventually result in the glass shattering into thumbnail-sized pieces. The glass is most susceptible to breakage due to damage to the edge of the glass, where the tensile stress is the greatest, but shattering can also occur in the event of a hard impact in the middle of the glass pane or if the impact is concentrated (for example, striking the glass with a hardened point).

Using tempered glass can pose a security risk in some situations because of the tendency of the glass to shatter completely upon hard impact rather than leaving shards in the window frame.

The surface of tempered glass does exhibit surface waves caused by contact with flattening rollers, if it has been formed using this process. This waviness is a significant problem in manufacturing of thin film solar cells. The float glass process can be used to provide low-distortion sheets with very flat and parallel surfaces as an alternative for different glazing applications.

Nickel sulfide defects can cause spontaneous breakage of tempered glass years after its manufacturing.

BEVELED GLASS

Beveled glass is a single pane of glass with a beveled edge. The term *beveled* refers to a cut made at an angle of less than 90 degrees. A beveled edge is typically used to add decorative style and has no functional purpose.

Beveled glass captures light in a unique way, creating a wide range of colors and enhancing the visual impact of the glass. Window and door manufacturers often use beveled glass to improve an simple design. The skillful arrangement of beveled glass, along with other decorative design elements increases both the visual appeal and the value of the final product.

You can often find examples of beveled glass in transom windows, door side lights and large, ornate front doors. The arrangement of multiple pieces of beveled glass within an entrance way creates an image of sophistication and style. This type of treatment has been consistently popular in North American architectural design since the late 1950's.

Although beveled glass is traditionally plain cut glass, there has been a trend to use this technique on colored textured glass to enhance a particular design. Typically, textured glass is 1/8" (0.32 cm) thick and has a very specific visual impact. The use of textured glass with a beveled edge increases the options available for designers to create new and interesting glass pieces.

HEBRON GLASS

Hebron Glass refers to glass produced in Hebron as part of a flourishing art industry established in the city during Roman rule in Palestine. Hebron's Old City still contains a quarter named the "Glass-Blower Quarter" and Hebron glass continues to serve as a tourist attraction for the city.

Traditionally, the glass was melted using local raw materials, including sand from neighbouring villages, sodium carbonate (from the Dead Sea), and coloring additives such as iron oxide and copper oxide. Nowadays, recycled glass is often used instead. Glass production in Hebron is a family trade, the secrets of which have been preserved and passed down by a few Palestinian families who operate the glass factories located just outside the city. The products made include glass jewellery, such as beads, bracelets, and rings, as well as stained glass windows, and glass lamps. However, due to the Palestinian-Israeli conflict, glass production has suffered a decline.

Production

Carefully moving molten glass as part of the modern production process.

Hebron glass was traditionally produced using sand from the village of Bani Na'im, east of Hebron, and sodium carbonate taken from the Dead Sea. Instead of sand, recycled glass is the primary raw material used to make Hebron glass today.

The precise production process is a trade secret maintained by the few Palestinian families who run the factories which continue to produce Hebron glass today, passed through generations by apprenticing children. As one master of the craft said, "You can learn to play the 'oud at any age, but unless you begin (glasswork) as a child, you will never become a master".

According to the Holy Land Handicraft Cooperative Society, the blowing technique employed is the same as was used by the ancient Phoenicians, though archaeologists and historians of glass agree that glassblowing was not common until the last few centuries BCE. Molten glass is withdrawn from a furnace on the end of an iron pipe, which is blown into as a metal tool called a *kammasha* is used to shape the glass. It is returned to the furnace and reshaped by the same process before being detached from the pipe and placed into a cooling chamber.

Jewellery

Glass beads for jewellery have traditionally been made in Hebron. Blue beads and glass beads with 'eyes' (owayneh) were made for use as amulets since they were considered particularly effective against the evil-eye. In the Museum of Mankind collections, there exist several glass necklaces that were made in Hebron during the Mandate period or earlier. Besides necklaces made of blue and green beads, and 'eyes' beads, there are examples of beads of small hands, also called a Hamsa, representing the hand of Fatimah, the daughter of the prophet Muhammad. Most of a woman's jewellery was given to her at marriage; in the early 1920s, in Bayt Dajan, a glass bracelet (ghwayshat) made in Hebron would be considered a necessary part of the jewellery of a bride's trousseau.

The jewellery store shows evil eye beads and Hebron-made glass bracelets sold alongside the shopkeeper's main ware of silver or metal wire. Photo taken 1900-1920 by American Colony, Jerusalem.

Hebron Trade Beads

In 1799, English traveller William George Browne mentioned the production of "Coarse glass beads...called Hersh and Munjir" in Palestine; The "Munjir" (*Mongur*) were large beads, while the Hersh (*Harish*) were smaller. These Hebron glass beads were used for trade, and export primarily to Africa from the early to mid-19th century. Spread throughout West Africa, in Kano, Nigeria, they were grounded on the edges to make round beads fit together on a strand more suitably. There, they picked up the name "Kano Beads", although they were not originally produced in Kano. By the 1930s, their value had decreased; in 1937, A. J. Arkell recorded the beads being sold "for a song" by Sudanese women to Hausa traders in Dafur.

SODA-LIME GLASS

Soda-lime glass is the most common form of glass produced. It is composed of about 70 percent silica (silicon dioxide), 15 percent soda (sodium oxide), and 9 percent lime (calcium oxide), with much smaller amounts of various other compounds. The soda serves as a flux to lower the temperature at which the silica melts, and the lime acts as a stabilizer for the silica. Soda-lime glass is inexpensive, chemically stable, reasonably hard, and extremely workable. These qualities make it suitable for manufacturing a wide array of glass products, including light bulbs, windowpanes, bottles, and art objects.

Soda-lime glass was produced throughout much of Europe for hundreds of years. Silica, in the form of sand, and limestone were abundant nearly everywhere. Soda ash was readily obtained from hardwood forests, though Venetian glassmakers favoured potash produced by burning seaweed.

HEATABLE GLASS

Electrically heatable glass and windows are relatively new products, which help solve problems in the design of buildings and vehicles.

The idea of heating glass is based on the use of energy-efficient low-emissive glass, which is generally simple silicate glass with a special metallic oxides coating. Low-emissive coating decreases heat loss by approximately 30%. Heatable glass can be used in all kinds of standard glazing systems, whether wood, plastic, aluminum or steel.

Heatable glass based on low-emissive coatings was first produced in high volume in the early 1980s. Today, heating glass is used in the construction of many kinds of buildings and in mass production of vehicles, ships and trains.

Heatable glass removes discomfort and other disadvantages induced by the low heat-insulating features of silicate glass. The effect of "cold glass" disappears when the surface of the glass is heated. Condensation is eliminated, along with ice and snow covering. The window's heat losses are compensated and room comfort is improved.

Heatable glass can be used as the principal system of heating and can be combined with floor heating and ceiling heating. Such combination helps reduce the total rate of heat loss of the building, thereby lowering heating expenses. Also, the active area of the room can be used more efficiently, as massive window-sill radiators are not needed. Initially, heating glass was produced by sputtering ordinary glass, and stable quality could not be guaranteed. A technological breakthrough took place in 1989 when the mass production of low-emissive glass began. The glass was coated during the manufacturing process.

Standard Windows
Window Construction

Windows play a significant role in room comfort. As a result, the area of glazing of buildings is constantly being increased. Window technologies are always progressing and it is common today to use low-emissive glass. Despite this progress, the low surface temperature of the glass is still a problem with constructive glazing. Heatable glass helps to solve this problem and increase the level of comfort in the room significantly. Heatable glass can be used in practically all kinds of glazing systems made of wood, plastic or aluminium.

Heatable glass and multiple glass panes can be used both in blind and openable constructions. Multiple glass panes made of heating glass can have one or two chambers. The advantages of multiple glass panes are their hermiticity and ability to decrease heat transfer significantly.

Optical Transmission and Heat Losses of Windows

If the temperature in the building is higher than the temperature outside, then heat leaks through the elements of construction. Windows are usually the elements of the buildings most vulnerable to heat loss. The heat loss though window constructions is about 20–25% of total heat loss.

Heat insulation of translucent constructions can be improved by increasing the number of glasses and chambers of multiple glass panes, but will result in increased construction cost and decreased optical transmission. The reasonable alternative is the usage of low-emissivity glass, which is practically the same as ordinary glass in terms of optical transmission, but it also reflects the heat radiation back into the room.

The major indicator which characterizes the ability of glass to reflect heat radiation is its emissivity (E) or the "emission factor". The emission factor of ordinary glass is

0.83; the factor for low-emissive glass can reach 0.03, so that more than 90% of accumulated heat will be reflected back into the room. The lower the emission factor is, the more effective is the material to reflect the heat, and the more heat it will accumulate. To compare, the emission factor of a multiple glass pane with two chambers, which is made of ordinary glass, is the same as the emission factor of a multiple glass pane with one chamber, which was produced with usage of low-emissive glass.

Besides energy-efficient functions in cold seasons of the year, low-emissive glass possesses the ability to reflect the excess of outside heat energy in summer seasons; the optical transmission coefficient is affected insignificantly in this way.

The additional factor of reduce of heat transfer of multiple glass panes is the usage of low thermal conductivity gases – Ar or Kr – to fill the chambers. In present-day multiple glass panes Ar is more often used, which helps to reduce heat losses by 10–20%, though the cost of multiple glass panes is insignificantly increased.

Influence of Window Surface Temperature on Comfort

There are two reasons why people feel discomfort when they are close to a cold window surface. First, a cold window is the reason for outflow of heat, which is produced by the cutaneous covering of the individual. Second, a cold window provokes the circulation of air, which is felt like a draft.

In order to reduce these factors the heating radiators are always placed under window sills. As far as people can feel cold and heat, the actual temperature of environment is not the only factor which defines the total level of comfort. In reality, the heat radiation of surrounding surfaces has a greater influence than air temperature. If the window surface is cold then to maintain the comfort atmosphere it is necessary to increase the heating temperature, but it will also increase energy consumption.

The problem of cold window can be solved effectively with the help of heatable glass. These windows allow to maintain the optimal comfort level and temperature of the room. The air temperature can be decreased at least by 1 degree if the temperature of surrounding surfaces has the same significance. You also do not have to install heating radiators and free the additional space for that. Besides, when turned off the multiple glass panes made of heatable glass act like ordinary low-emissive glass.

Multiple Glass Panes made of Heating Glass

Structure of Heatable Glass

The idea of heatable glass is based on usage of energy-efficient low-emissive glass, where the coating plays the role of heating element. It can be used both in production of multiple glass panes and as a part of triplex, which has also the function of protective glazing.

The technological process of production of multiple glass panes made of heatable glass is practically the same as the process of ordinary multiple glass panes production. The main difference is the presence of power supply and, if necessary, temperature sensor. The temperature sensor allows to track the temperature of heating glass and eliminates the possibility of overheating of the product.

In order to prevent shocks, the conductive coating is always placed inside the multiple glass pane or laminated unit. Only safe tempered glass, the strength of which is a lot higher than the strength of ordinary glass, is used in production of heatable glass. When the hardened glass is destroyed there are safe splittings. Also the current-carrying coating loses its integrity and the automatic fuse, which turns off the power supply of the glass, is activated. The electrodes are placed inside the lamination and no one can reach them without destruction of the product.

Usage of Heatable Glass

Alarm Glass.

Transport Glass.

Heating glass is mostly used for heating of windows. It is especially useful for rooms where people spend much time by the windows, at home or at work. The most common usage of heating glass—windows of cottages, office buildings and also big areas—leaded panes, translucent roofing, garret windows, canopies and so on.

Heating glass is used for defogging and prevention of frosting of windows of pools, saunas and other buildings of such kind. Insofar as heatable glass has a current-carrying coating, it can be used as the sensor of alarm systems. When the glass is destroyed the system of protection is activated and it results in activation of alarm system. This kind of product is widely used on objects of tightened standards in questions of protection: nuclear power plants, stations of air navigation control, museums, special storehouses, etc.

Heatable glass is also used in production of windows for different kinds of vehicles: electric and diesel locomotives, vessels and boats, various kinds of aircraft and automobiles. One of well-known examples of application of heating glass is armored windows, because the protective glazing is very thick and is disposed to frosting. The usage of heating glass is especially urgent in terms of being the part of armored multiple glass of Smart Glass of switchable transparency, because the heating significantly decreases the period of reaction of liquid crystals structure.

The power consumed by products depends on the type of use. Power of about 50–100 watts per square meter of the window is generally enough for maintenance of comfort temperature in the room and for maintenance of glass surface temperature at the rate from +20 degrees to +30 degrees. When the heatable glass is used as the only source of heat it is necessary to maintain the glass surface temperature at the rate from +30 degrees to –45 degrees and provide the power of 100 to 300 watt for 1 square meter of the window. The power needed for vehicle windows reaches 1.5 kilowatts per square meter or more, which is why there are such tight standards in terms of sputtering of current-carrying components.

Heat power of about 500–700 watts per square meter of glazing is necessary for snow unloading and taking ice-covering off the outside protective translucent constructions in low temperatures and windy environments.

Technology of Production

Heatable glass is produced by lamination of two or more sheets of silicate glass. The most widespread technologies are the following technologies of panel production according to the type of materials used:

- EVA v ethylene-vinyl-acetate film with good adhesion to glass. Major advantages: low cost of both film and equipment. One needs only a primitive furnace with vacuum bags for production. Disadvantages: high rate of opalescence, especially after multicoat lamination, with the lapse of time yellowness appears. EVA has low shear strength, especially in low temperatures; it results in delamination (layering).

- PVB – Polyvinyl butyral film with high rate of adhesion to glass. Major advantages: low cost of mass production of laminated glass, insignificant rate of opalescence, high quality of product. Disadvantages: high initial cost of equipment, it is necessary to have autoclave, press for preliminary hot pressing, "clean" room, and qualified personnel. Besides that, triplex made with the help of PVB technology can not be use in wet environment.

- TPU – thermoplastic polyurethane film with very high rate of adhesion to glass. Major advantages: insignificant rate of opalescence, insensible to humidity, mechanical effects and severe atmosphere; very high quality of the product. Disadvantages: high cost of film and equipment, it is necessary to have autoclave, "clean" room, and qualified personnel.

- Photocurable polymers (resins) – so-called "filling technology". Major advantages: low cost of both resin and equipment. Only an ultraviolet furnace and a minimum of additional equipment is needed to produce laminated glass. Disadvantages: it is necessary to have qualified personnel for the work. Heating laminated glass that is produced with the help of this technology is insensible to humidity and temperature influence, has high shear strength.

PLATE GLASS

Plate glass, flat glass or sheet glass is a type of glass, initially produced in plane form, commonly used for windows, glass doors, transparent walls, and windscreens. For modern architectural and automotive applications, the flat glass is sometimes bent after production of the plane sheet. Flat glass stands in contrast to *container glass* (used for bottles, jars, cups) and *glass fibre* (used for thermal insulation, in fibreglass composites, and optical communication).

Flat glass has a higher magnesium oxide and sodium oxide content than container glass, and a lower silica, calcium oxide, and aluminium oxide content. (From the lower soluble oxide content comes the better chemical durability of container glass against water, which is required especially for storage of beverages and food).

Most flat glass is soda–lime glass, produced by the float glass process. Other processes for making flat glass include:

- Rolling (rolled plate glass, figure rolled glass).
- Overflow downdraw method.
- Blown plate method.
- Broad sheet method.
- Window crown glass technique.
- Cylinder blown sheet method.
- Fourcault process.
- Machine drawn cylinder sheet method.
- Plate polishing.

Quality and Damage

Scratches can occur on sheet glass from accidental causes. In glass trade terminology these include "block reek" produced in polishing, "runner-cut" or "over/under grind" caused by edge grinding, or a "sleek" or hairline scratch, as well as "crush" or "rub" on the surface.

BOROSILICATE GLASS

Borosilicate is a type of glass that was at first developed in the late 19th century in Germany by Otto Schott. Borosilicate is a special type of glass that consists of silica and boron oxide. This is a glass that is commonly used for its very advantageous properties like low coefficient of thermal expansion, that is, it is more resistant to thermal shock when compared to common glass. This is the reason why this type of glass is most preferred for making laboratory glassware, as it does not react with the chemicals nor is it affected by thermal changes. This is an advantageous quality as this means that this glass, when subjected to extreme stress, will rather crack or break into large pieces instead of shattering into many small ones.

Also, this glass is less than dense than the usual soda-lime glass and with a relatively low refractive index. Due to these various properties, this glass is most preferred for many uses in the glass industry. Besides laboratory glassware, borosilicate glassware is also used to make microwave glass cookware, due to its resistance to heat. This is the reason it is also used to make high quality beverage glassware, as it increases the durability and also the dishwasher compatibility.

Borosilicate glass is also used in making aquarium heaters, flashlights, guitar slides, high intensity discharge lamps, etc. Even astronomical reflecting telescopes use borosilicate glass due to its low coefficient of expansion with heat. Besides these, borosilicate glass also has its use in the semiconductor industry and in the making of thermal insulation tiles of the famed Space Shuttle! Due to its high resistance and relatively inert nature, borosilicate glasses are also used for immobilisation and disposal of radioactive waste.

TIFFANY GLASS

Tiffany glass refers to the many and varied types of glass developed and produced from 1878 to 1933 at the Tiffany Studios in New York, by Louis Comfort Tiffany and a team of other designers, including Frederick Wilson and Clara Driscoll.

In 1865, Tiffany traveled to Europe, and in London he visited the Victoria and Albert Museum, whose extensive collection of Roman and Syrian glass made a deep impression on him. He admired the coloration of medieval glass and was convinced that the quality of contemporary glass could be improved upon. In his own words, the "Rich tones are due in part to the use of pot metal full of impurities, and in part to the uneven thickness of the glass, but still more because the glass maker of that day abstained from the use of paint".

Tiffany was an interior designer, and in 1878 his interest turned towards the creation of stained glass, when he opened his own studio and glass foundry because he was unable to find the types of glass that he desired in interior decoration. His inventiveness both as a designer of windows and as a producer of the material with which to create them was to become renowned. Tiffany wanted the glass itself to transmit texture and rich colors and he developed a type of glass he called "Favrile".

The glass was manufactured at the Tiffany factory located at 96-18 43rd Avenue in the Corona section of Queens from 1901 to 1932.

Types

Opalescent Glass

Opalescent glass.

The term "opalescent glass" is commonly used to describe glass where more than one color is present, being fused during the manufacture, as against *flashed* glass in which two colors may be laminated, or silver stained glass where a solution of silver nitrate is superficially applied, turning red glass to orange and blue glass to green. Some opalescent glass was used by several stained glass studios in England from the 1860s and 1870s onwards, notably Heaton, Butler and Bayne. Its use became increasingly common. Opalescent glass is the basis for the range of glasses created by Tiffany.

Favrile Glass

Tiffany patented Favrile glass in 1892. Favrile glass often has a distinctive characteristic that is common in some glass from Classical antiquity: it possesses a superficial iridescence. This iridescence causes the surface to shimmer, but also causes a degree of opacity. This iridescent effect of the glass was obtained by mixing different colors of glass together while hot.

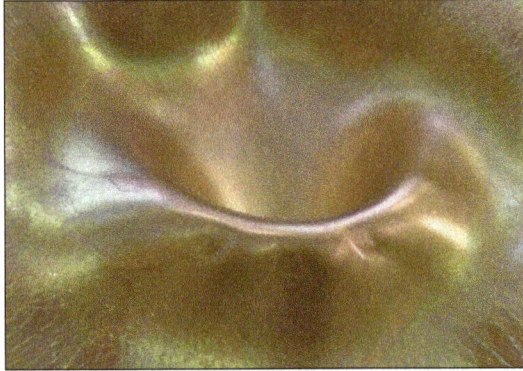

Favrile glass.

Streamer Glass

Streamer glass refers to a sheet of glass with a pattern of glass strings affixed to its surface. Tiffany made use of such textured glass to represent, for example, twigs, branches and grass.

Streamer glass.

Streamers are prepared from very hot molten glass, gathered at the end of a punty (pontil) that is rapidly swung back and forth and stretched into long, thin strings that rapidly cool and harden. These hand-stretched streamers are pressed on the molten surface of sheet glass during the rolling process, and become permanently fused.

Fracture Glass

Fracture glass refers to a sheet of glass with a pattern of irregularly shaped, thin glass wafers affixed to its surface. Tiffany made use of such textured glass to represent, for example, foliage seen from a distance.

Fracture glass.

The irregular glass wafers, called *fractures*, are prepared from very hot, colored molten glass, gathered at the end of a blowpipe. A large bubble is forcefully blown until the walls of the bubble rapidly stretch, cool and harden. The resulting glass bubble has paper-thin walls and is immediately shattered into shards. These hand blown shards are pressed on the surface of the molten glass sheet during the rolling process, to which they become permanently fused.

Fracture-streamer Glass

Streamer-fracture glass.

Fracture-streamer glass refers to a sheet of glass with a pattern of glass strings, and irregularly shaped, thin glass wafers, affixed to its surface. Tiffany made use of such textured glass to represent, for example, twigs, branches and grass, and distant foliage.

The process is as above except that both streamers and fractures are applied to sheet glass during the rolling process.

Ring Mottle Glass

Ring mottle glass refers to sheet glass with a pronounced mottle created by localized, heat-treated opacification and crystal-growth dynamics. Ring mottle glass was invented by Tiffany in the early 20th century. Tiffany's distinctive style exploited glass containing a variety of motifs such as those found in ring mottle glass, and he relied minimally on painted details.

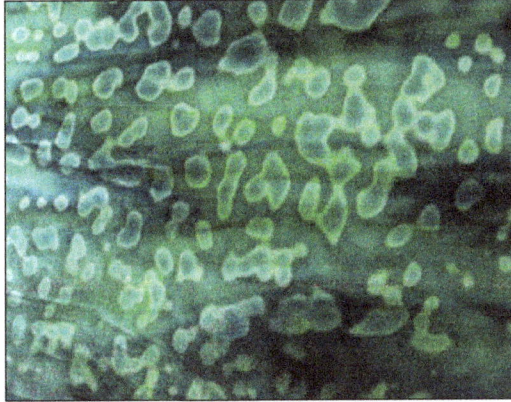

Ring mottle glass.

When Tiffany Studio closed in 1928, the secret formula for making ring mottle glass was forgotten and lost. Ring mottle glass was re-discovered in the late sixties by Eric Lovell of Uroboros Glass. Traditionally used for organic details on leaves and other natural elements, ring mottles also find a place in contemporary work when abstract patterns are desired.

Drapery Glass

Drapery glass refers to a sheet of heavily folded glass that suggests fabric folds. Tiffany made abundant use of drapery glass in ecclesiastical stained glass windows to add a 3-dimensional effect to flowing robes and angel wings, and to imitate the natural coarseness of magnolia petals.

Drapery glass.

The making of drapery glass requires skill and experience. A small diameter hand-held roller is manipulated forcefully over a sheet of molten glass to produce heavy ripples, while folding and creasing the entire sheet. The ripples become rigid and permanent as the glass cools. Each sheet produced from this artisanal process is unique.

Cutting Techniques

In order to cut streamer, fracture or ripple glass, the sheet may be scored on the side without streamers, fractures or ripples with a carbide glass cutter, and broken at the score line with breaker-grozier pliers.

In order to cut drapery glass, the sheet may be placed on styrofoam, scored with a carbide glass cutter, and broken at the score line with breaker-grozier pliers, but a band-saw or ringsaw are the preferred.

STAINED GLASS

Stained glass is the coloured glass used for making decorative windows and other objects through which light passes. Strictly speaking, all coloured glass is "stained," or coloured by the addition of various metallic oxides while it is in a molten state. Nevertheless, the term stained glass has come to refer primarily to the glass employed in making ornamental or pictorial windows. The singular colour harmonies of the stained-glass window are less due to any special glass-colouring technique itself than to the exploitation of certain properties of transmitted light and the light-adaptive behaviour of human vision. Rarely equalled and never surpassed, the great stained-glass windows of the 12th and early 13th centuries actually predate significant technical advances in the glassmaker's craft by more than half a century. And much as these advances undoubtedly contributed to the delicacy and refinement of the stained glass of the later Middle Ages, not only were they unable to arrest the decline of the art, but they may rather have hastened it to the extent that they tempted the stained-glass artist to vie with the fresco and easel painter in the naturalistic rendition of their subjects.

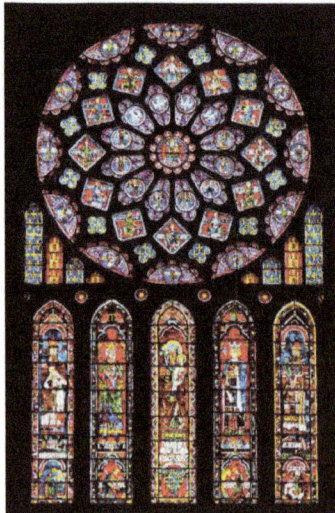

Chartres Cathedral: stained-glass rose window.

Neither painting on stained glass nor its assembly with grooved strips of leading is an indispensable feature of the art. Indeed, the leaded window may well have been preceded by windows employing wooden or other forms of assembly such as the cement tracery that has long been traditional in Islamic architecture, and the single most important technical innovation in 20th-century stained glass, slab glass and concrete, was a variation on the earlier masonry technique.

Coronation of Edward VI, stained glass, Mansion House, London.

Elements and Principles of Design

Of all the painter's arts, stained glass is probably the most intractable. It is bound not only by the many light-modulating factors that affect its appearance but also by comparatively cumbersome, purely structural demands. And yet no other art seems so little earthbound, so alive, so intrinsically beguiling in its effect. This is because stained glass, far more directly and intensively than other media, exploits the interaction between two highly dynamic phenomena, the one physical and the other organic. The physical factor is light and all of the myriad changes in the general light level and the location and intensity of particular light sources that occur as a matter of course not only from moment to moment but from place to place—a prairie to a forest, a greenhouse to a dungeon. The other phenomenon is the spontaneous light-adaptive process of vision, which seeks to maintain orientation in all luminous environments.

Architecture, by determining the apparent brightness value of the light seen through its window openings, always establishes a definite scale of brightness values with which the stained-glass artist must work. Because the light that penetrated the interior of the 12th- and early 13th-century church took on a brilliance, even harshness, in contrast to the surrounding darkness, the artisans of the period logically composed their windows with a palette of deep, rich colours. When for doctrinal or economic reasons only clear glass could be used, it was decorated with a fine opaque mesh of grisaille, or monochromatically painted ornament, that effectively broke up and softened the light. Later, as the walls of the churches were opened up to admit more and more light, the difference between the interior and exterior light levels was no longer great enough to illuminate the dense, saturated rubies and blues of the earlier period.

In the 14th and 15th centuries, generally higher keyed, drier, and more muted colour harmonies were developed. This reflected a growing preference for lighter, less awesome effects and an actual limitation that the architecture of the time imposed upon the medium of stained glass.

The static elements of the glass and its architectural setting are modified by the element of change inherent in natural light. A seemingly endless spectrum of changes in the appearance of stained glass is a result of the changes in the intensity, disposition, atmospheric diffusion, and colour of natural daylight. The luminous life of stained glass, therefore, can best be observed by watching the organic effect of light on the window through the course of a day. If one were to enter Chartres Cathedral just after sunrise on the morning of a clear day, it would be to the east windows, especially those in the clerestory, that one's eyes would first be drawn. They alone will have come fully to life, and all of the others will still seem to half-exist in a kind of hushed twilight. Gradually, as the sun rises in the sky, these windows will become more luminous. Then the east windows will begin to lose their earlier brilliance to those all along the south flank of the cathedral, which by midday will be fairly aglow from the direct rays of the sun. The light streaming through the south windows, however, will have raised the light level inside the north windows opposite them sufficiently to create a distinct, though by no means unpleasant, muting of the radiance of the latter. If the sun at this point disappears behind a cloud and the sky becomes generally overcast, the appearance of all of the windows is immediately and dramatically altered. Because the light, now diffused, comes more or less equally from all directions, the south windows will lose some of their earlier brilliance and vivacity and the north windows will recover theirs. The overall atmosphere of the cathedral is distinctly cooler and graver in its effect, and more than ever before one begins to become aware of absolute differences in the tonality of the various windows themselves. The grisaille windows in the east end of the cathedral, the highly keyed 15th-century window in the Vendôme Chapel in the south aisle of the nave, and the three 12th-century windows over the great west portal all stand out as being substantially more luminous than the rest. If, late in the afternoon, the sun reappears, the viewer is treated to an extraordinary spectacle as the blues in the west windows, by far the most intense in the cathedral, are further emblazoned by the direct rays of the sun. Should the main doors of the cathedral be opened, the direct rays of the late afternoon sun, streaming halfway down the nave of the cathedral, will cast a blinding pall over all the windows within their vicinity until the doors are closed once more. Then as the sky begins to redden with the setting sun, the intense 12th-century blues in the west windows lose their former intensity, and the warmer colours, especially the rubies, become so fiery and assertive that they seem almost to have displaced the blues as the predominant colour in the windows. Finally, when the sun is gone the whole cathedral is plunged once more into a deep twilight, which gradually diminishes until there is no light at all.

Insofar as stained glass may be considered an art of painting, it must be considered an art of painting with light. Whatever techniques or materials it may employ, its own most unique and indispensable effects are always the product of colouring, refracting, obscuring, and fragmenting light.

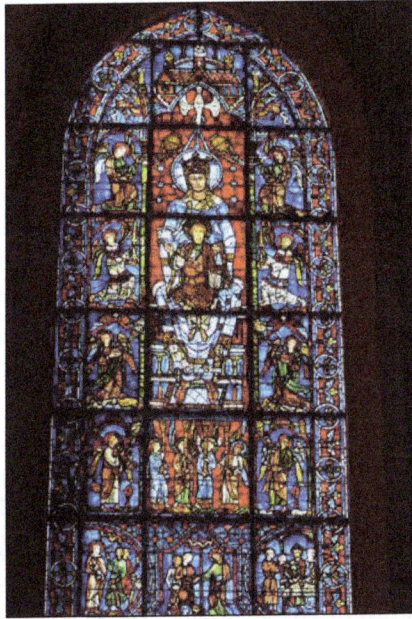

Chartres Cathedral: "Beautiful Window".

Materials and Techniques

Contrary to popular belief, the glassmaker and the stained-glass artist could seldom have been the same person even in the earliest times; in fact, the two arts were rarely practiced at the same location. The glassmaking works was most readily set up at the edge of a forest, where the tremendous quantities of firewood, ash, and sand that were necessary for the making of glass could be found, whereas the stained-glass-window-making studios were normally set up near the major building sites. The stained-glass artist, thus, has always been dependent upon the glassmaker for his primary material. Coloured with metallic oxides while in a molten state—copper for ruby, cobalt for blue, manganese for purple, antimony for yellow, iron for green—sheets of medieval glass were produced by blowing a bubble of glass, manipulating it into a tubular shape, cutting away the ends to form a cylinder, slitting the cylinder lengthwise down one side, and flattening it into a sheet while the glass was still red hot and in a pliable state. It was then allowed to cool very slowly in a kiln so that it would be properly annealed and not too difficult to cut up into whatever shapes might be required for the design. Since these sheets of glass, with the exception of a type known as flashed glass, were intrinsically coloured with one basic colour throughout, changes from one colour to another in the design of a window could be effected only by introducing separate pieces of glass in each of the requisite colours.

Whether by accident or by deliberate intent, the glass made in the 12th and 13th centuries had almost the ideal combination of crudity and refinement for stained glass.

The sheets, 10 by 12 inches (25 by 30 centimetres) in size, were both flat enough and thin enough to be cut very accurately into the necessary shapes, yet still variable enough in thickness (from less than $1/_8$ inch [3 millimetres] to as much as $5/_{16}$ inch [8 millimetres]) to have rich transitions in the depth of their colours. With the progress of glass technology in the Middle Ages and Renaissance came the ability to produce larger, thinner, and flatter sheets of glass in a considerably larger range of colours than had been possible in the 13th century. At each distinguishable stage in this development, however, the glass became less visually interesting as an aesthetic element in its own right. The Gothic Revivalists later recognized this effect, and in the mid-19th century they initiated a return to the earlier methods of producing glass. They developed the so-called "antique" glass, which is remarkably similar in colour, texture, and shading to the glass that was used in the 12th- and 13th-century windows. "Antique" glass remains the basic material used in stained-glass windows to this day.

Traditional Techniques

The art of stained glass is the translucent offspring of such earlier art forms as mosaic and enamelling. From the mosaicist came the conception of composing monumental images out of many separate pieces of coloured glass. Cloisonné enamelling probably inspired not only the technique of binding these pieces together with metal strips but that for treating the strips themselves as a positive design element. From the enamellers must also have come the near-black vitreous enamel made from rust powder and ground glass that was mixed with a mild water-based glue to form a paint. This could be used to render more or less opaquely onto glass the details of figures, ornaments, and inscriptions.

The technique of making stained-glass windows is first described in the Schedula diversarum artium, a compendium of craft information probably written between 1110 and 1140 by the monk Theophilus (tentatively identified as the 12th-century goldsmith Rugerus of Helmarshausen). First, a full-sized cartoon, or line drawing, of the window was painted directly onto the top of a whitewashed table, showing the division of the various colour areas into individual pieces of glass. Next, sheets of glass of the appropriate colours were selected and from these pieces were cut, or, more accurately, cracked away with a red hot iron. By applying the hot iron to the edge of the sheet it was possible to start a crack that could then be guided more or less in the direction in which the iron was moved, thus enabling the glazier to break away from the sheet of glass a piece of approximately the right shape and size. This he would then further shape by "grozing," or crumbling away bits of glass from its edges with a notched tool known as a grozing iron. When all of the pieces were thus accurately cut to shape, with due allowance between pieces for the leads that would join them together, the details of the design were painted onto the glass wherever necessary with vitreous enamel. The pieces were then placed in a kiln and fired at a temperature just hot enough to fuse the enamel to the glass. This done, the windows were ready for assembly with grooved strips of lead that look in cross section like the letter H. The glazier would begin by butting together on his workbench two long strips of lead, to form a corner of the panel. He would then set the corner piece of glass in place between these two leads and cut another strip of lead just long enough to surround the rest of the piece. Against this lead he would then be able to set the next piece of glass, and so on across the panel, until it was completely assembled on the glazing bench. The joints between the leads were then soldered, the panel was waterproofed by rubbing a putty compound under the leads, and it was ready for installation.

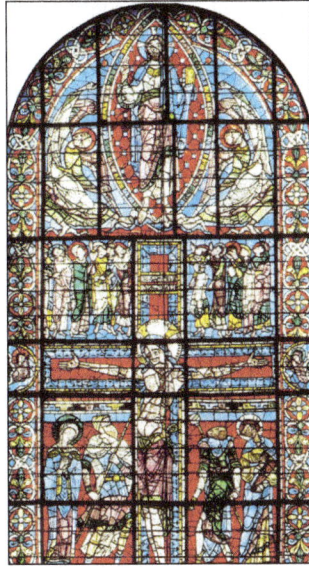

Poitiers Cathedral: Crucifixion window.

Because of the flexibility of the leading it was found necessary to divide all but the very smallest windows into a series of separate leaded panels and to insert iron framing members, or armatures, between the panels. In the earliest single-figure lancet windows, such as the Prophets in Augsburg Cathedral, the divisions tend to be purely functional. Very soon, however, more ambitious windows became much too large to be handled in this manner. Whereas the Augsburg Prophets measure only about 12 square feet (1.1 square metres) in area, the Poitiers Cathedral Crucifixion window contains approximately 175 square feet (16.3 square metres) of stained glass, and the Life of Christ in Chartres contains more than 250 square feet (23.2 square metres). A much more elaborate system of subdivisions in the window opening, consisting of vertical as well as horizontal members, was developed. These systems of supports often formed a geometric pattern that was incorporated in the overall design of the window. In fact, it was the ingenious conversion of this structural necessity into a positive design element that set the stage for the creation of the medallion windows of the great Gothic cathedrals. By utilizing these armatures to delineate the principal ornamental subdivisions of the windows, as in the Chartres Good Samaritan, the glass painters were able to fuse a complex didactic imagery and an austere architecture into one of the most compelling artistic unities of Western art. At the same time, particularly in the upper levels of a church, stone mullions began to be employed for the same purpose. The most spectacular examples are the great rose windows, in which masonry is so literally dissolved into fenestration, and the individual window opening so completely absorbed into the overall pattern, as to defy any meaningful distinction between window and wall. This perfect fusion of image, ornament, and structure, with each deriving strengths from the others that none would ever have alone, was one of the most significant turning points in the history of stained glass. From this point on the relation between stained glass and architecture begins to decline. The aims, techniques, and achievements of the stained-glass artist begin to resemble those of the fresco and easel painters, and it is by the standards applicable to the latter that the stained glass of the 14th, 15th, and 16th centuries must be judged.

Developments in the 14th Century

The first significant developments in the glass painter's craft appear to have been made more or less simultaneously in the early years of the 14th century. Glass in a range of previously unavailable secondary colours—smoky ambers, moss greens, and violet—becomes generally available for the first time. The technique of staining glass yellow by painting it with silver salts is discovered. The glass painters also begin to develop a number of techniques for shading or modelling forms with vitreous enamel by applying translucent matts of halftone to the whole surface of the window and delicately brushing it away where highlights are desired. Darker shading is sometimes reinforced by painting on the outer as well as the inner surface of the glass. The uses of line also become increasingly refined and versatile, especially in the 15th century.

To these refinements of the craft was added one wholly new technique, the abrasion of flashed glass. Ruby glass, whose unique composition made this technique possible, was a laminated glass, although it appears to be coloured intrinsically throughout like all of the other glass in the early windows. Because the metallic agent used to produce its colour was so dense, all but the thinnest films of ruby were opaque. To obtain sufficient translucency, either the glassmaker had to suspend striations of ruby in a clear glass, thereby creating the "streaky rubies" of the early 13th century, or the glass was "flashed"; that is, clear glass while still pliant was dipped into molten coloured glass, thus coating its surface with a thin film of colour. Detailed effects, unhindered by intricate leading, could then be achieved by grinding away portions of this coloured film, first on ruby glass and then on other colours deliberately "flashed" for this purpose. To these colours could now also be added the silver salts stain in tones of yellow ranging from the palest canary tint to a deep fiery amber, depending on how heavily the stain was applied and how thoroughly it was fired. The whole gamut of more or less translucent tonalities that could be created with vitreous enamel were also used. Taken altogether, these techniques when used in combination represented a considerable liberation of stained glass from what was increasingly considered to be the "tyranny" of the lead line.

The technique of grinding flashed glass was first practiced in the late 13th and early 14th centuries; one of the earliest extant examples is in the church at Mussy-sur-Seine in France, where the windows have a blue groundwork covered all over, or diapered, with ruby roses with white centres, each rose being a single piece of glass. This type of work, however, was not common until the 15th and 16th centuries.

Later Developments

At the end of the 15th century a whole new range of vitreous enamels was developed, and by the middle of the 16th century the technique of painting in enamel colours on glass began to be of major importance. In this method, granulated coloured glass of the desired colour is mixed with a flux of clear ground glass and fired onto the surface of the glass. Enamel painting was not altogether successful either technically or aesthetically, since the colours thus created were translucent rather than transparent, generally pallid, and of uncertain durability. Political disturbances in the mid-17th century created a scarcity of coloured glass throughout Europe, and gradually the traditional use of coloured glass was replaced by the new technique.

Between the 16th and 20th centuries the developments in the craft of making stained-glass windows were purely utilitarian. In the 16th century the diamond glass cutter was invented, and in the 18th century hydrofluoric acid was introduced as a means of etching flashed glass. In the 19th and 20th centuries, gas and electric kilns and soldering irons were used, as were plate-glass easels upon which stained-glass panels could be temporarily mounted for painting before they were leaded. The largest palette of glass—the widest range of colours, textures, and thicknesses that the art has ever known—was also developed in the 20th century. Contemporary technical innovations include the slab glass and concrete windows developed in France about 1930, where glass set in concrete provides an alternative to leading. In the mid-20th century such experimental techniques as bonding glass to glass with transparent resin glues were developed. Measured purely by technical standards, contemporary stained glass has never been rivalled in its versatility as an instrument of artistic expression.

Subject Matter

In the Middle Ages ecclesiastical art was primarily didactic. The subjects painted in the windows played an important part in the expounding of the Scriptures and the glorification of the church and its saints.

The iconographic program of medieval stained-glass windows for ecclesiastical buildings is a product of several factors. To begin with, the cruciform plan of the churches themselves created four focal areas. Each area, by its architectural form and orientation to the sun, tended to elicit the development of certain subjects or types of subject. In Chartres, for example, the five central windows of the choir clerestory and the north rose window are consecrated to the Virgin, the south rose window to the glorification of Christ, and the west rose window to the Last Judgment. In Bourges Cathedral the huge figures of the Apostles in the south clerestory are paired off against the prophets in the north clerestory, the representatives of the New Testament thereby receiving the full light of the sun and their Old Testament counterparts the more crepuscular light of the north sky. The great rose windows, whose circular form is itself cosmological in its implications, are invariably devoted to cosmological themes: the Last Judgment, the Apocalypse, the cycle of the year as expressed in the signs of the zodiac, the glorification of Christ and of the Virgin as the rulers of heaven. On the other hand, one of the reasons that the theme of the Jesse tree remained popular throughout the Middle Ages was that it lent itself to such a rich variety of ornamental treatments. And finally there was the will of the donors of the windows, whose personal preferences determined the subjects of many excellent works that clearly cannot be related to any comprehensive iconographic program. Some idea of the scope of these medieval enterprises can be indicated by the fact that Chartres, by no means the largest of the cathedrals, contains more than 27,000 square feet (2,500 square metres) of stained glass, in 176 windows. Of the 64 windows on the lower level, all but a few are medallion windows, which contain anywhere from 20 to 30 or more separate pictorial compositions; and the three rose windows, each more than 40 feet (12 metres) in diameter, are vast composite creations. The work of at least nine separate master designers has been distinguished in the windows of the cathedral, which was completely glazed in less than 40 years, between about 1203 and 1240.

It must be assumed that clerics supplied the master glazier with a program to which he had to conform. A 12th-century manuscript in the British Museum contains a series of circular drawings illustrating the life of St. Guthlac. These drawings might have been intended as a model for a glazier, but the scenes could equally well have been expressed in wall paintings, sculpture, or metalwork. There is more complete knowledge for the later Middle Ages. The glazier was given written instructions from which to prepare provisional sketches that were submitted for the patron's approval before being redrawn in actual size to form the final cartoon. The provisional sketch was known as a vidimus. One example of such written instructions is the program for a window given by Henry VII to the Grey Friars Church at Greenwich, England.

There is ample evidence to show that by the 14th century it was the practice of glaziers to have a stock of finished cartoons, executed on parchment or paper, which could be adapted for different glazing schemes. That these cartoons were used and reused over a long period can be deduced from the will of a York glazier, who died in 1450, in which he bequeathed to his son all his cartoons.

It is evident that in the later Middle Ages the master glazier's workshop was a highly organized enterprise, capable of producing various classes of designs, according to the expense his patrons were prepared to incur. Although the donor, cleric or layman, exercised considerable influence over the choice of subject and its manner of representation, the finished design was essentially the creation of the master glazier. The latter was often an artist in his own right, expressing in the formal language of his own technique the artistic aspirations of his time.

References

- Safety-glass, technology: britannica.com, Retrieved 10 May, 2019

- Soda-lime-glass, technology: britannica.com, Retrieved 28 August, 2019

- An-overview-of-borosilicate-glass: borosil.com, Retrieved 25 June, 2019

- Stained-glass, art: britannica.com, Retrieved 05 March, 2019

- He, Fei (December 2009). "Rapid fabrication of optical volume gratings in Foturan glass by femtosecond laser micromachining". Applied Physics A. 97 (4): 853–857. doi:10.1007/s00339-009-5338-4

Processes of Glass Preparation

The processing of glass consists of various important manufacturing steps such as glass melting, glass batch calculation, calculation of glass properties, glass coating, annealing of glass and glass inspection. This chapter closely examines these key stages of glass manufacturing to provide an extensive understanding of the subject.

GLASS MELTING

Before melting can begin, raw materials – or "batches" – must be gathered together. The first step in this process involves weighing – so the ingredients of the so-called batch recipe are measured. The batch consists of the specified ingredients. Precision is especially important here, since even just 1 extra gram can completely change the result. In order to ensure the batch is uniform, all ingredients are thoroughly mixed in large mixing units. The subsequent glass melting phase is a central step in glass production. Ultimately, it is the raw material composition and quality of raw materials, the manner of heat applied, and the specific melting process that determine the type of melted glass and thus the end product.

Crucible ovens or pot furnaces were originally used for this melting process. Here the individual steps in the glass melting process are performed in sequential order. With tank melting, which is the most common method used today, the processes are completed continuously in sequential order as they move through the system. Both melting processes are still used today for glass melting.

GLASS BATCH CALCULATION

Glass batch calculation or glass batching is used to determine the correct mix of raw materials (batch) for a glass melt.

Principle

The raw materials mixture for glass melting is termed "batch". The batch must be measured properly to achieve a given, desired glass formulation. This batch calculation is based on the common linear regression equation:

$$N_B = (B^T \cdot B)^{-1} \cdot B^T \cdot N_G$$

with N_B and N_G being the molarities 1-column matrices of the batch and glass components respectively, and B being the batching matrix. The symbol "T" stands for the matrix transpose operation, "-1" indicates matrix inversion, and the sign "·" means the scalar product. From the molarities matrices N, percentages by weight (wt%) can easily be derived using the appropriate molar masses.

Example Calculation

An example batch calculation may be demonstrated here. The desired glass composition in wt% is: 67 SiO_2, 12 Na_2O, 10 CaO, 5 Al_2O_3, 1 K_2O, 2 MgO, 3 B_2O_3, and as raw materials are used sand, trona, lime, albite, orthoclase, dolomite, and borax. The formulas and molar masses of the glass and batch components are listed in the following table:

Formula of glass component	Desired concentration of glass component, wt%	Molar mass of glass component, g/mol	Batch component	Formula of batch component	Molar mass of batch component, g/mol
SiO_2	67	60.0843	Sand	SiO_2	60.0843
Na_2O	12	61.9789	Trona	$Na_3H(CO_3)_2{*}2H_2O$	226.0262
CaO	10	56.0774	Lime	$CaCO_3$	100.0872
Al_2O_3	5	101.9613	Albite	$Na_2O{*}Al_2O_3{*}6SiO_2$	524.4460
K_2O	1	94.1960	Orthoclase	$K_2O{*}Al_2O_3{*}6SiO_2$	556.6631
MgO	2	40.3044	Dolomite	$MgCa(CO_3)_2$	184.4014
B_2O_3	3	69.6202	Borax	$Na_2B_4O_7{*}10H_2O$	381.3721

The batching matrix B indicates the relation of the molarity in the batch (columns) and in the glass (rows). For example, the batch component SiO_2 adds 1 mol SiO_2 to the glass, therefore, the intersection of the first column and row shows "1". Trona adds 1.5 mol Na_2O to the glass; albite adds 6 mol SiO_2, 1 mol Na_2O, and 1 mol Al_2O_3, and so on. For the example given above, the complete batching matrix is listed below. The molarity matrix N_G of the glass is simply determined by dividing the desired wt% concentrations by the appropriate molar masses, e.g., for SiO_2 67/60.0843 = 1.1151.

$$
B = \begin{bmatrix}
1 & 0 & 0 & 6 & 6 & 0 & 0 \\
0 & 1.5 & 0 & 1 & 0 & 0 & 1 \\
0 & 0 & 1 & 0 & 0 & 1 & 0 \\
0 & 0 & 0 & 1 & 1 & 0 & 0 \\
0 & 0 & 0 & 0 & 1 & 0 & 0 \\
0 & 0 & 0 & 0 & 0 & 1 & 0 \\
0 & 0 & 0 & 0 & 0 & 0 & 2
\end{bmatrix}
\qquad
N_G = \begin{bmatrix}
1.1151 \\
0.1936 \\
0.1783 \\
0.0490 \\
0.0106 \\
0.0496 \\
0.0431
\end{bmatrix}
$$

The resulting molarity matrix of the batch, N_B, is given here. After multiplication with the appropriate molar masses of the batch ingredients one obtains the batch mass fraction matrix M_B:

$$N_B = \begin{bmatrix} 0.82087 \\ 0.08910 \\ 0.12870 \\ 0.03842 \\ 0.01062 \\ 0.04962 \\ 0.02155 \end{bmatrix} \quad M_B = \begin{bmatrix} 49.321 \\ 20.138 \\ 12.881 \\ 20.150 \\ 5.910 \\ 9.150 \\ 8.217 \end{bmatrix} \quad \text{or} \quad M_B(100\% \text{normalized}) = \begin{bmatrix} 39.216 \\ 16.012 \\ 10.242 \\ 16.022 \\ 4.699 \\ 7.276 \\ 6.533 \end{bmatrix}$$

The matrix M_B, normalized to sum up to 100% as seen above, contains the final batch composition in wt%: 39.216 sand, 16.012 trona, 10.242 lime, 16.022 albite, 4.699 orthoclase, 7.276 dolomite, 6.533 borax. If this batch is melted to a glass, the desired composition given above is obtained. During glass melting, carbon dioxide (from trona, lime, dolomite) and water (from trona, borax) evaporate.

Simple glass batch calculation can be found at the website of the University of Washington.

Advanced Batch Calculation by Optimization

If the number of glass and batch components is not equal, if it is impossible to exactly obtain the desired glass composition using the selected batch ingredients, or if the matrix equation is not soluble for other reasons (i.e., the rows/columns are linearly dependent), the batch composition must be determined by optimization techniques.

CALCULATION OF GLASS PROPERTIES

The calculation of glass properties (glass modeling) is used to predict glass properties of interest or glass behavior under certain conditions (e.g., during production) without experimental investigation, based on past data and experience, with the intention to save time, material, financial, and environmental resources, or to gain scientific insight. It was first practised at the end of the 19th century by A. Winkelmann and O. Schott. The combination of several glass models together with other relevant functions can be used for optimization and six sigma procedures. In the form of statistical analysis glass modeling can aid with accreditation of new data, experimental procedures, and measurement institutions (glass laboratories).

Global Models

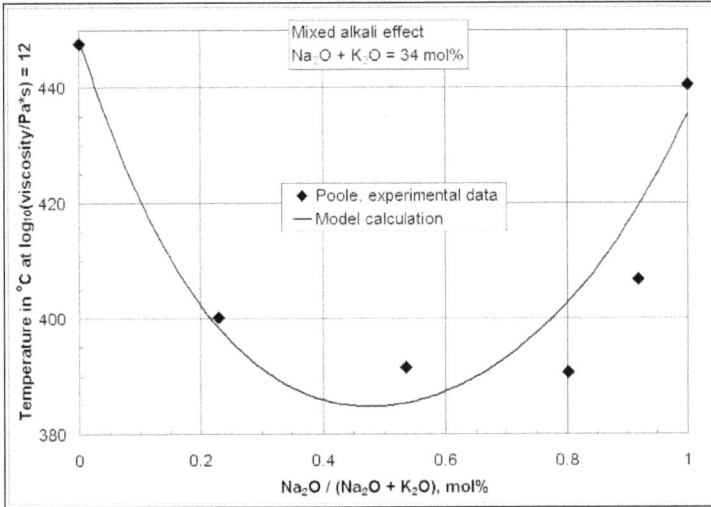

The mixed-alkali effect: If a glass contains more than one alkali oxide,
some properties show non-additive behavior. The image shows, that
the viscosity of a glass is significantly decreased.

Schott and many scientists and engineers afterwards applied the additivity principle to experimental data measured in their own laboratory within sufficiently narrow composition ranges (local glass models). This is most convenient because disagreements between laboratories and non-linear glass component interactions do not need to be considered. In the course of several decades of systematic glass research thousands of glass compositions were studied, resulting in millions of published glass properties, collected in glass databases. This huge pool of experimental data was not investigated as a whole, until researchers published their global glass models, using various approaches. In contrast to the models by Schott the global models consider many independent data sources, making the model estimates more reliable. In addition, global models can reveal and quantify non-additive influences of certain glass component combinations on the properties, such as the mixed-alkali effect as seen in the adjacent diagram, or the boron anomaly. Global models also reflect interesting developments of glass property measurement accuracy, e.g., a decreasing accuracy of experimental data in modern scientific literature for some glass properties, shown in the diagram. They can be used for accreditation of new data, experimental procedures, and measurement institutions (glass laboratories). In the following sections (except melting enthalpy) empirical modeling techniques are presented, which seem to be a successful way for handling huge amounts of experimental data. The resulting models are applied in contemporary engineering and research for the calculation of glass properties.

Non-empirical (*deductive*) glass models exist. They are often not created to obtain reliable glass property predictions in the first place (except melting enthalpy), but to establish relations among several properties (e.g. atomic radius, atomic mass, chemical bond strength and angles, chemical valency, heat capacity) to gain scientific insight. In future, the investigation of property relations in deductive models may ultimately lead to reliable predictions for all desired properties, provided the property relations are well understood and all required experimental data are available.

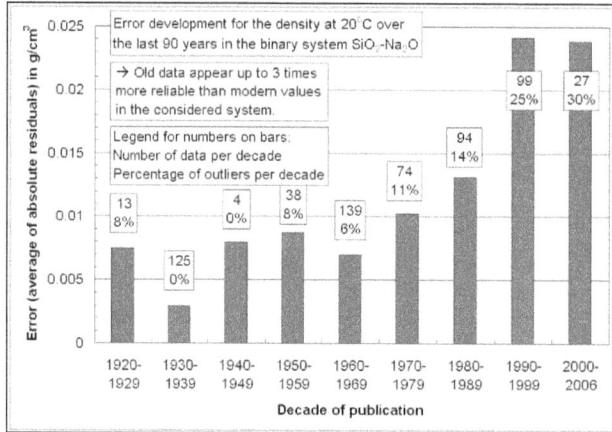

Decreasing accuracy of modern glass literature data for the density at 20 °C in the binary system SiO_2-Na_2O.

Methods

Glass properties and glass behavior during production can be calculated through statistical analysis of glass databases such as GE-SYSTEM SciGlass and Interglad, sometimes combined with the finite element method. For estimating the melting enthalpy thermodynamic databases are used.

Linear Regression

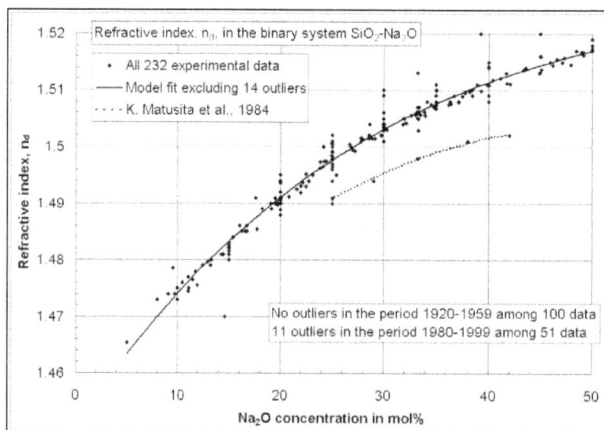

Refractive index in the system SiO_2-Na_2O. Dummy variables can be used to quantify systematic differences of whole dataseries from one investigator.

If the desired glass property is not related to crystallization (e.g., liquidus temperature) or phase separation, linear regression can be applied using common polynomial functions up to the third degree. Below is an example equation of the second degree. The C-values are the glass component concentrations like Na_2O or CaO in percent or other fractions, the b-values are coefficients, and n is the total number of glass components. The glass main component silica (SiO_2) is excluded in the equation below because of over-parametrization due to the constraint that all components sum up to 100%. Many terms in the equation below can be neglected based on correlation and significance analysis.

Non-linear Regression

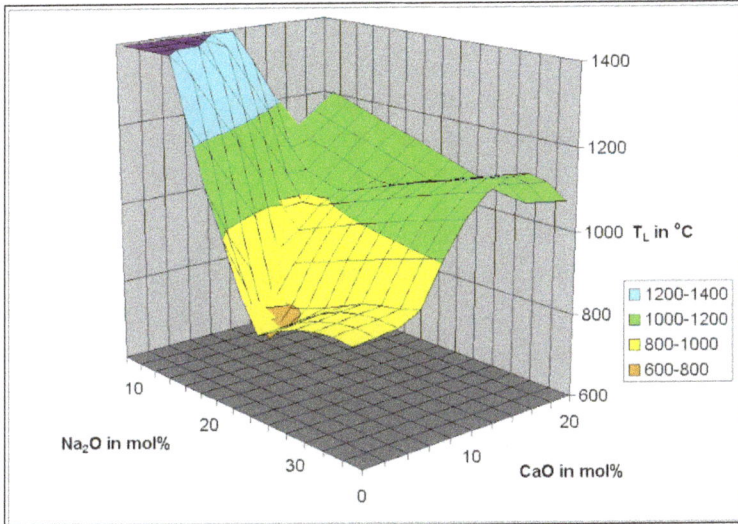

Liquidus surface in the system SiO_2-Na_2O-CaO using disconnected peak functions based on 237 experimental datasets from 28 investigators. Error = 15 °C.

The liquidus temperature has been modeled by non-linear regression using neural networks and disconnected peak functions. The disconnected peak functions approach is based on the observation that within one primary crystalline phase field linear regression can be applied and at eutectic points sudden changes occur.

Glass Melting Enthalpy

The glass melting enthalpy reflects the amount of energy needed to convert the mix of raw materials (batch) to a melt glass. It depends on the batch and glass compositions, on the efficiency of the furnace and heat regeneration systems, the average residence time of the glass in the furnace, and many other factors.

Finite Element Method

For modeling of the glass flow in a glass melting furnace the finite element method is applied commercially, based on data or models for viscosity, density, thermal conductivity, heat capacity, absorption spectra, and other relevant properties of the glass melt. The finite element method may also be applied to glass forming processes.

Optimization

It is often required to optimize several glass properties simultaneously, including production costs. This can be performed, e.g., by simplex search, or in a spreadsheet as follows:

1. Listing of the desired properties;

2. Entering of models for the reliable calculation of properties based on the glass composition, including a formula for estimating the production costs;

3. Calculation of the squares of the differences (errors) between desired and calculated properties;

4. Reduction of the sum of square errors using the Solver option in Microsoft Excel with the glass components as variables. Other software (e.g. Microcal Origin) can also be used to perform these optimizations.

It is possible to weight the desired properties differently. The combination of several glass models together with further relevant technological and financial functions can be used in six sigma optimization.

GLASS COATING

Glass coatings consist of particles of silicon dioxide (SiO_2) that work to repel contamination and increase shine. The SiO_2 means the coating contains "pure glass" with the glass in a molecular form, held in a liquid. The coating does not become "glass" until it is allowed to dry.

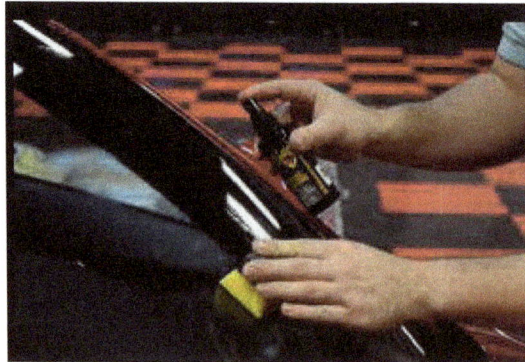

Whilst a glass coating is about 500 times thinner than a human hair, it leaves a wet look that is popular in the car industry, even though the coating is invisible to the naked eye.

Glass coating has been around for a while and is still popular in the car market. It is marketed under several names including liquid glass, nano-glass, ceramic glass, and quartz coating.

The two major types of glass coating are quartz silane and silica. Quartz silane is more expensive because the manufacturing process is more difficult, but it provides a very durable and extra shiny coating. In contrast, silica is less durable, making it cheaper than a quartz silane coating. A glass coating will form a hard, durable and semi-permanent shield over paint, glass and wheels, with different coatings available for each surface.

Purpose of Glass Coating

Glass coating once applied is an invisible, durable coating that repels water, is anti-static, and protects the surface from dirt, dust and other pollutants sticking to it. This makes cleaning much easier. Glass coating also increases resistance to scratches by up to 10 times the resistance of non-coated materials.

Glass coating can be used on a number of different surfaces, not just vehicles, including tiles, glass, bowls, grout, floors and pools and is safe to use on glass, ceramics, stainless steel, acrylics, painted surfaces and wheels.

With vehicles, a glass coating improves visibility during rain on windshield glass. Glass coating is available in brands that are either colourless or a pale yellow. There is only a mild smell with this water-based coating.

Glass coating is much simpler to apply than ceramic coating since it does not have to be applied at a specific heat like a ceramic coating. Also, glass coating has a less rigid application process meaning there is more time available to polish and buff the coating to get the result you want. This means there is less risk of making an error, but also means a correction is not so costly. A coat can last anything from six months to one year under normal conditions, when it can simply be reapplied.

Types of Glass Coating

There are several types of glass coatings, which by their very nature do not include the need for heat, including:

- High-gloss water-repellent glass coating: A two-component glass coating that produces instant luminosity. A special chemical reaction creates a long-lasting water-repellent effect and prevents adhesion of raindrops and water spots.

- Hybrid Titanium/Glass Coating: Titanium enriched glass coating generates a hard membrane with water repellent properties that protects the paint surface from dirt, acid, UV rays and scratches.

- Pure Glass Coating: A pure glass coating is high in silica making it an extremely durable and hard glass coating that reflects light with specific wavelengths, producing an instant luminous shine and gloss. It also works as an excellent water repellent with additional scratch protection.

- Nano-titanium glass coating: For a very deep shine and spectacular reflection on the paint a nano-titanium glass coating also provides a harder protective layer against wear and the ageing of the vehicles clearcoat.

- Advanced water-repellent glass coating: Based on a three-dimensional molecular framework it forms a highly durable clear glass coat with strong water repelling effects on the paint and produces a unique "candy-like" gloss.

- Heat resistant hard glass coating for wheels and body: A hard glass coating with a tight molecular structure that creates long-lasting protection resistant against heat of up to 1,300°C. Therefore, it is particularly efficient in protecting wheels from burning-hot brake dust.

ANNEALING OF GLASS

If a hot glass object is cooled "too quickly," it may be strained at room temperature, and therefore may break easily. For small, or thin-walled shapes (particularly those made of glasses having low expansions) the effect may not be serious. For more massive pieces, the strain can be very serious. The amount of strain (observed in a polariscope) depends upon how quickly the object passes through a critical temperature range. The range depends on the composition of the glass but is usually about 450°C. If the glass is cooled slowly through that range, so that the temperature near the surface is never very different from that of the interior, then the strain in the resulting object is much reduced. Such glass is said to be *annealed*.

Glasses differ from crystalline solids in that glasses do not have distinct melting points. This difference is explained by the fact that the chemical bonds holding the atoms together in a regular crystalline structure are identical. When the crystalline solid is heated, all the bonds break at exactly the same temperature. Below this temperature, called the melting point, the material is solid; above the melting point the material is a liquid.

In contrast, the bonds in glass show a range of strengths. When a glass is heated, these bonds break over a range of temperatures. As a result, a glass softens gradually as it heats. It is convenient to describe the behavior of glass with temperature in terms of *viscosity*. Viscosity is the resistance of a liquid to flow. The unit of viscosity is the *poise*.

- Water has a low viscosity of 0.01 poise.

- SAE 30 motor oil is 1.0 poise.

- The viscosity of glass at room temperature cannot be measured.

Working point: The viscosity at which glass is suitable for working or forming is 10^4 poises.

Softening point: The temperature at which a glass fiber less than one millimeter in diameter will stretch under its own weight at a rate of one millimeter per minute when suspended vertically. This occurs at a viscosity of $10^{7.6}$ poises.

Annealing point: The temperature at which strain in thin sections will be removed in 15 minutes and which viscosity is 10^{13} poises. The strain point is that temperature below which permanent strain cannot be introduced.

Annealing: The process of slowly cooling a completed object in an auxiliary part of the glass furnace, or in a separate furnace. This is an integral part of glassmaking because if a hot glass object is allowed to cool too quickly, it will be highly strained by the time it reaches room temperature; indeed, it may break as it cools. Highly strained glasses break easily if subjected to mechanical or thermal shock.

Poise: A CGS absolute unit of viscosity that is equal to one dyne-second per square centimeter.

Viscous: A term usually applied to liquids, and means in a qualitative sense, the resistance that a liquid offers to flow; molasses has a high viscosity. Viscosities are expressed in a unit called the poise. The viscosity of water at room temperature is .010 poise: of SAE 30 motor oil is about 1.0 poise. The viscosity of most glasses at room temperature is about 10^{19}-10^{22} poises, which is about as high a viscosity as can be measured. Viscosity is related to temperature.

GLASS INSPECTION

Quality and reliability are determined by the design, operation and control of the process from the delivery of the raw materials to the output of the final product. Despite the control which is exercised in modern glass manufacturing operations, defects can occur which make the product unacceptable. Many of these defects cannot be detected at the stages of manufacture where they arise, and, as such, it is essential, if customer satisfaction is to be assured, that they are detected in the final product, and only those products of acceptable quality despatched to the customer. The costs of inspection, the losses associated with rejection for scrap, and the reworking of products which can be made acceptable are an integral part of the total product cost. These costs can only be reduced by reducing reject losses. It is important, therefore, that, in addition to rejecting defective products, the inspection system also provides the information needed by production in carrying out the preventive action necessary to achieve and maintain a consistent and acceptable level of product quality.

To determine the optimum balance between inspection costs, preventive action and reject losses is a basic management problem. Facts on these costs, as they affect manufacturing costs, are relatively easy to obtain. Facts on their value, as they affect customer reaction, are more difficult to ascertain, as the loss of goodwill caused by the despatch of below-standard products can have a delayed action such as the eventual loss of orders. Assuming a satisfactory and consistent level of inspection, the relative economics between reject losses and preventive action are illustrated in figure, to the left of the optimum, the losses due to defectives are greater than the costs of preventive action. To the right of the optimum, the costs are uneconomical due to perfectionism, and the costs of prevention are greater than the losses due to rejects.

Essential Features of a Product Quality Inspection System

It is evident from the foregoing that a product quality inspection system must produce, at minimum added product cost, the information required for:

- Feedback to manufacturing to assist in the manufacture of products of consistent and acceptable quality.

- Feed forward of the information necessary to ensure that products which fail to meet this quality are rejected for correction or scrap.

Further, if the inspection system is to be technically and cost effective, then it is essential that:

- The product quality standards necessary to ensure customer satisfaction are clearly defined.

- The methods and frequency of inspection are compatible with, and related to the desired final product quality, and to the nature and frequency of occurrence of the quality defects.

- The information obtained is presented at the right time and in a form necessary to ensure its effective utilization.

- The actions necessary to ensure corrective action and rejection of unacceptable products are clearly defined and implemented.

- The recommended and applied actions are periodically checked to verify that they have achieved their alms.

The Case for Automatic Inspection

Because of the transparent properties of glass, principal requirements are that it should have an acceptable level of optical perfection and freedom from defects. Due to this optical requirement, inspection has traditionally been carried out by human inspectors. The combination of eye and brain has a sensitivity of judgement which is not always technically and economically easy to replace by automatic inspection. However, manufacturing processes are becoming more highly mechanised and automated, and in high throughput situations, visual inspection presents a speed on response limitation. The sub jective nature of human inspection also imposes a limitation on the ability to measure defect severity; the efficiency of inspection depending on the experience and skill of the particular examiner. The combination of time limitation and subjective judgement frequently results in the unnecessary rejection of acceptable products and acceptance of those which are unsatisfactory. This, in turn, leads to increased reject losses and customer complaints. In addition, at high throughputs the examiner is unable to accurately record a continuous count of the type, severity and frequency of occurrence of defects as required by production to reduce quality defects, and improve the yield of acceptable products.

The limitations of human inspection can be overcome by automatic methods. The methods can, however, introduce problems of their own. For example, they may not be able to identify the type of defect, or to relate optical imperfections, such as distortion, to human judgement of acceptability. Despite the potential problems which may arise, techniques for automatic inspection exist which are capable of operating at the required level of efficiency, and of providing the information necessary to achieve a satisfactory and acceptable level of quality in products at minimum cost.

Problems and Examples of Techniques of Automatic Inspection

A first essential of any automatic inspection system is a clear definition of product quality acceptable to the customer. Examples of the principal defects which can occur in glass manufacturing operations and the categories into which they can be placed are given in Table.

The problems associated with the automatic examination and detection of these defects, and the techniques which have been developed are as follows:

Discrete Defects

In flat glass production and, in particular, where the ribbon of glass leaving the lehr and entering the warehouse is immediately cut by mechanised cutting operations into the sizes required by the customer, it is important that discrete defects are located and assessed to the agreed quality standards so that below-standard glass can be immediately identified and rejected during the subsequent cutting and handling operations.

Discrete defects may be extremely small and can be located anywhere within the ribbon of glass being produced. Because of their size and random distribution, they are amongst the most difficult faults to locate. With on-line inspection, the time available for inspection will decrease with the rate of production. For example, for a float glass ribbon 3.048 m. (120 ins.) wide, travelling at 914.4 m./s. (600 ins./min.), 55.8 m² (600 ft.²) of glass will pass the inspection zone in 1 minute. These speeds are beyond the capability of the human examiner, and if automatic inspection is to be used, then because of the random distribution of the defects, every unit area of the glass must be examined in the time available. This means that some system of scanning the surface of the glass must be developed, and if all the defects are to be found, the area scanned at any particular instance of time must be of the same order as the smallest fault to be located. Therefore, that the problem of their detection is immense. Once they are located, however, they have the advantage that, in the case of flat glass, the acceptance or rejection of the defects listed in Table for any particular application of the product is determined by their size.

Numerous systems have been developed for the automatic inspection of flat glass and, in particular, float glass. The systems developed fall into three categories, namely:

Flying Spot Systems

In this case, the surface of the glass is continuously scanned by a spot of light, the source of which may be either a laser or incandescent lamp. The scan is achieved by reflecting the incident beam from a rotating mirror system, and the light, after passing through the glass, is brought to a focus on a photo detector. The spot size and the speed of scan are such that adjacent line scanning is obtained on the surface of the glass.

Image Dissection Systems

Here the ribbon is illuminated from below and an image of the glass is formed in the plane of a rotating disc positioned in the opposite side of the glass. The disc incorporates a series of radial slits, and as it rotates, the image is repeatedly scanned. The speed of rotation of the disc is such that adjacent line scanning again results as the ribbon of glass passes through the field of view, and successive images are formed in the plane of the scanning disc.

Electro-optical Techniques

In this system, a collimated light source is directed through the ribbon onto a bank of photo-transistors located transversely across the ribbon of glass. An aperture is positioned in front of the detectors such that the presence of a discrete defect causes a change in light intensity reaching the detector.

In all the systems, the presence of a defect results in a change in light intensity reaching the detector. The change in light intensity may be used simply to indicate the presence of a defect, or the length of the defect in the direction of travel of the ribbon. In the case of the scanning systems, this is obtained by counting the number of times the defect is scanned as it passes through the field of view. As the majority of defects are either circular or elongated in the direction of draw of the ribbon of glass , and since it is the maximum dimension which is generally important, this single measurement is usually adequate. If required, however, a width measurement could be obtained from the pulse width produced.

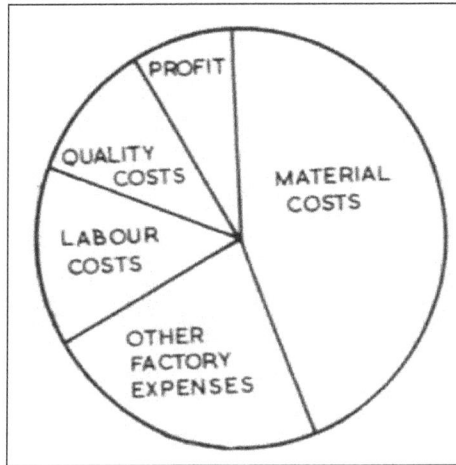

Constituents of Product Costs.

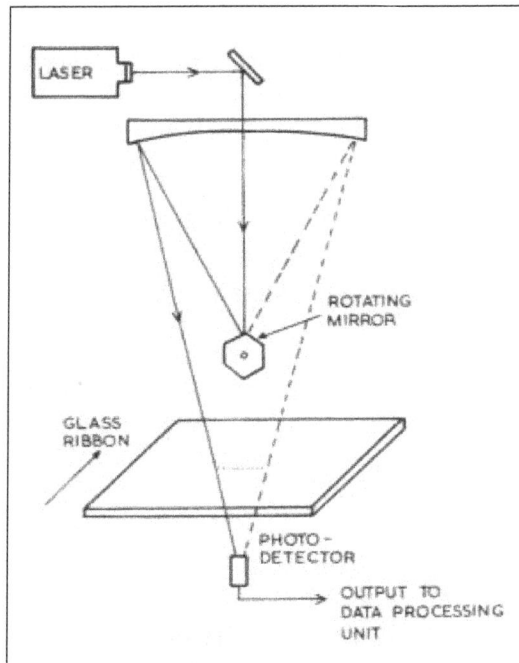

Schematic Diagram of Flying Spot Scanner.

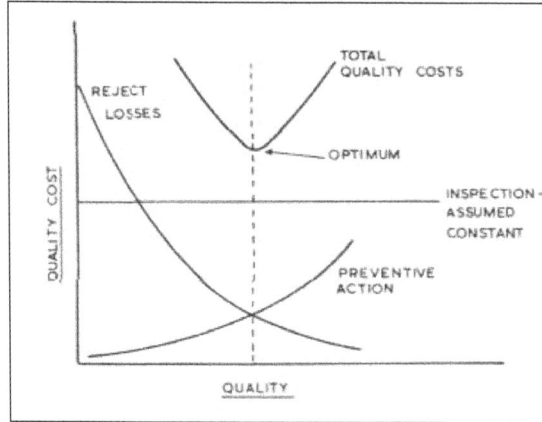

Economics of Inspection Costs Reject Losses and Preventive Action.

Table: Categories and Examples of Manufacturing Defects.

Examples of Defects	Category
Stones , bubbles, ream knots	Discrete defects
Size , thickness, shape , contour	Dimensional defects
Refractive index changes (ream). Changes Ln thickness and surf ace contour.	Distortion defects
Scratches , scars , cracks , crizzles	Surface defects

Having located and sized the defects, the information can then be used for:

- Automatic feedback to manufacturing (process control) of their location, occurrence and size.

- Comparison of the defect size with pre-set standards (customer requirements).

- Selection of fault-free, full width plates of glass in the lengths required to meet the order book.

- Marking on the surface of the ribbon the position of defects which fail to meet agreed quality standards so that, when the ribbon is cut to smaller sizes , the cut plates containing unacceptable defects can be rejected for cutting into smaller fault- free sizes.

- Control of the position of the cutting heads so that the pattern of cutting can be automatically adjusted to cut round the defects and produce defect-free sizes.

A typical installation is shown diagrammatically in figure below. Since all scanners so far developed have a limited length of scan, a number of scanners are required to cover the full width of the ribbon. The scanner information is processed by a computer which prints out information for process control, and, at the same time, directs information on defect location to a marker system downstream of the scanners. The correct time delay to the firing of the appropriate marker is obtained from a rotary pulse generator which measures the speed of the glass.

To mark the position of the defect across the width of the ribbon to a positional accuracy of ± 12 .5 mm. (± 2 in.) requires 120 markers for a 3.048 m. (120 in.) wide ribbon.

Dimensional Defects

The need to inspect for dimensions depends on the product, the method of forming and fabrication, and the final application.

In flat glass production, size and squareness of cut plates are dependent on the design and operation of the cutting machine. Since this is capable of great stability of operation, no more than periodic checks are required to ensure that the system is correctly adjusted and the plate dimensions are within tolerance.

Glass thickness, however, is dependent on the distribution of temperature and hence viscosity across the width of the ribbon. Although an y thickness changes which occur only take place slowly, it can be of advantage to have a continuous record of thickness in order to:

- Indicate trends so that the forming operation can be corrected to ensure that the thickness remains within tolerance.

- Continuously monitor thickness during a planned change to ensure that the thickness is brought to the new value as soon as possible.

The principal methods which are employed are radioisotope and optical gauges.

In the radioisotope gauge, the thickness is determined by the absorption of the incident beam as it passes through the glass.

In optical gauges, the thickness is measured by focusing a narrow beam of light onto the surface of the glass, and measuring the displacement between the beams reflected from the top and bottom surfaces. Errors which would otherwise be introduced by wedge in the glass, that is, nonparallelism of the surfaces, can be overcome by imaging the surfaces of the glass onto the photo detector. The systems have the advantage that they are non-contacting and therefore do not mark the surface.

In pressed glass operations, such as the production of ophthalmic blanks and particularly colour television (C.T.V.) face panels, contour of the surface is of prime importance.

In C.T.V. the inside surface of the panel must be within 0.00059 in. (0.015 mm.) of the true contour. This is necessary in order to ensure that the space between the inner surface of the panel and the shadow mask is maintained as uniform as possible, otherwise a change in spacing will cause the electron beam passing through the shadow mask to land on other than the correct contour dot with which it is intended to register.

Manual inspection using dial gauges is obviously slow, tedious to perform, and liable to error. To overcome these disadvantages, and to ensure the quickest possible awareness of drifts in tolerance so that rapid corrective action can be taken, automatic inspection systems, have been developed whereby the panel is accurately located on reference supports, and the contour, with reference to these supports, measured by a plurality of pneumatic or linear voltage transducers. The outputs from the transducers are displayed on a panel of lights to give immediate visual indication of out of tolerance situations for rejection of below-standard products. Equally, the information can be progressed through a computer for statistical analysis and print-out of deviations and their location for feedback to production for process correction.

Distortion Defects

Distortion in flat and curved glass is caused by the deviation of light as it passes through the thickness, due to either changes in refractive index or changes in thickness and contour.

Distortion can be quantified in terms of the deviation produced in the path of a beam of light passing through the thickness of the glass. The eventual customer, however, in many instances assesses distortion on a subjective basis, and formulates his judgement on the appearance of objects as viewed by the eye through an area of the glass. Automatic inspection, on the other hand, assesses distortion on a point-to-point basis across the surface of the glass. The main difficulty, therefore, lies in establishing a relationship between this point-to-point information and the subjective area analysis by the inspector or customer.

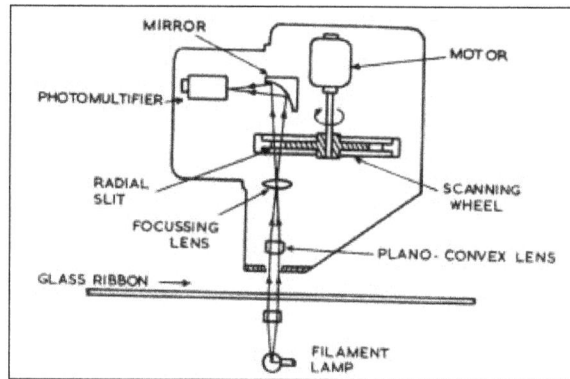

General Arrangement of Image Dissection Scanner.

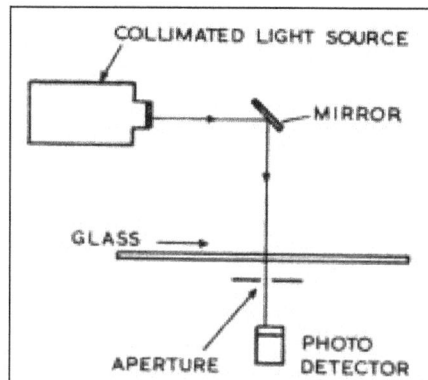

Electro-optical Defect Detector.

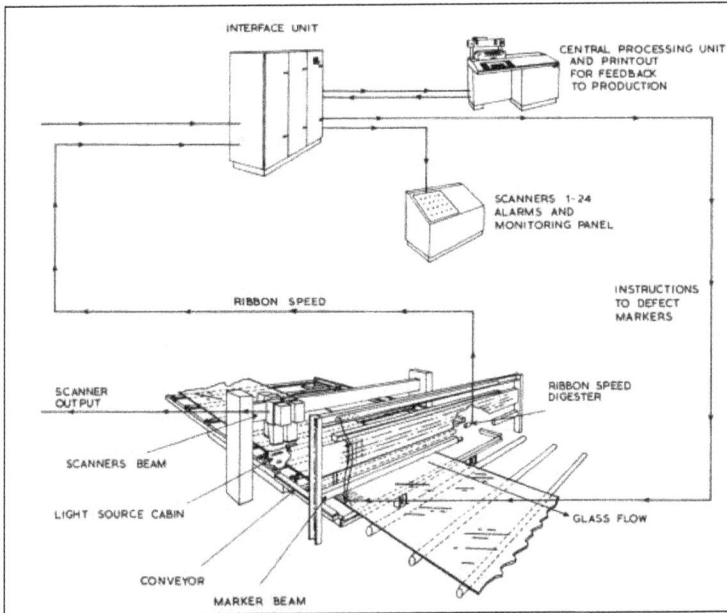

Typical Installation of Discrete Defect Scanner System.

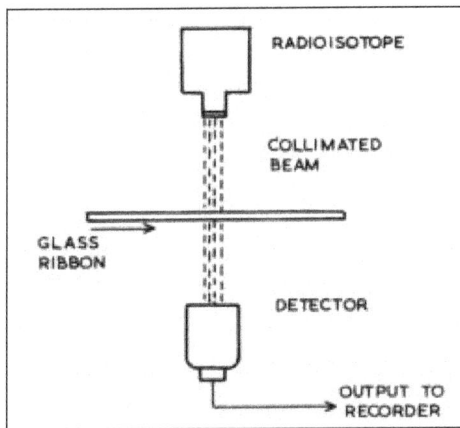

Radioisotope Thickness Gauge.

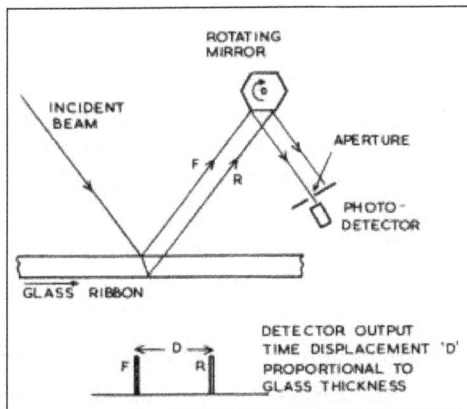

Principle of Operation of Optical Thickness Gauge.

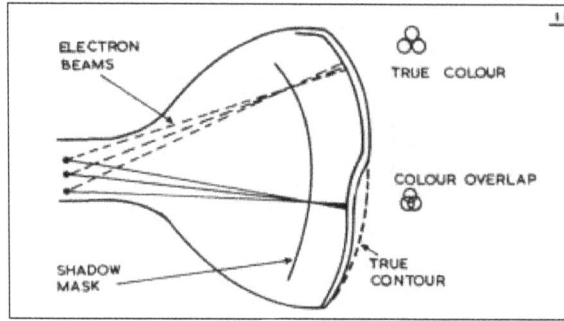

Effect of Panel Contour on Colour Register.

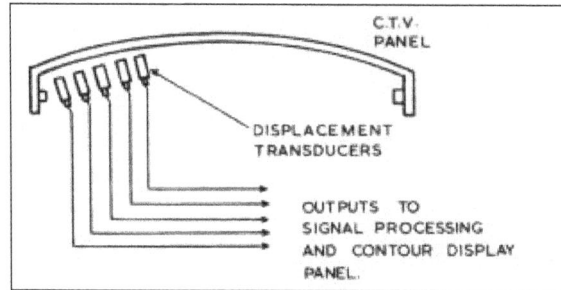

Schematic of Television Panel Contour Measurement.

Many techniques and data processing systems have been developed to examine and quantify distortion. In the majority of cases, they are used as an aid to the production, and control outgoing quality by identifying trends and regions of gross distortion; the final decision on acceptability being made by the examiner.

Two of the main methods which have been developed for automatic distortion measurement are the single-pass image projection system and the double-pass 'Schlieren' method. Both methods are basically extensions of the Foucault knife-edge test in which the optics are designed to produce a "perfect" image of an object. By introducing an optically imperfect medium, the image is deformed by an amount directly proportional to the deformations in the sample. The information is usually analysed in one dimension by means of a single-stop (knife-edge) or double-stop (rectangular aperture), or in two dimensions by a circular aperture, to produce a quantitative measurement of both the position and degree of the angular deviation of the sample. In automatic inspection systems, the deformation results in a change of light intensity on the detector, the output from which is a measure of the deviation introduced by the glass. By electrically differentiating the angular deviation signal, the value equivalent to the refractive power of the glass can also be obtained. To provide total coverage of the glass requires multiple scans and is therefore time consuming in such cases as for the inspection of automobile windscreens where 100% inspection of the total area is required.

In flat glass production, distortion defects, when they occur, are of a more periodic and regular nature, and the distortion can be assessed by continuously tracking an analyser to and fro across the surface of the glass.

Surface Defects

In flat glass, the acceptance or rejection of surface defects such a s scratches is dependent on their severity and on the particular application of the glass. For example, for silvering purposes, the glass must be free of surface scratches, as once the glass is silvered, their appearance becomes more pronounced and objectionable and the product has to be rejected. This is costly, both in terms of the reduction in yield and the cost of production. The scratches, because of their size, are most difficult to detect, and because of their random location and orientation, it has not, so far, been possible to develop automatic means of detection.

In container ware, cracks and crizzles are of a more coarse nature, and their occurrence results in rejection of the particular article. With the increase in manufacturing speeds and the introduction of automatic packaging, the automatic detection of these defects is essential, and a great number of methods have been developed and applied to this problem. To detect these defects, bottles entering the inspection zone are rotated at high speed such that the presence of a crack passing through a light beam incident on the area under examination causes scattering of the incident light which can then be detected by an appropriately located photo detector. The output of the photo detector can then be used to operate an automatic rejection device. By employing more than one light source and detector, selected parts of the container can be simultaneously examined for defects of this type.

References

- Melting, advanced-optics-capabilities: schott.com, Retrieved 08 February, 2019

- Glass-coating: cori-coatings.be, Retrieved 15 July, 2019

- Annealing-glass: cmog.org, Retrieved 04 April, 2019

- Terese Vascott; Thomas P. Seward III (2005). High Temperature Glass Melt Property Database for Process Modeling. Wiley-American Ceramic Society. ISBN 1-57498-225-7

- Terese Vascott; Thomas P. Seward III (2005). High Temperature Glass Melt Property Database for Process Modeling. Wiley-American Ceramic Society. ISBN 1-57498-225-7

Glass Production Techniques

The glass production techniques can be classified into commercial techniques and artistic techniques. Commercial techniques include glassblowing, glass casting, vitrification, etc. and artistic techniques include caneworking, glass fusing, slumping, etc. All these diverse glass production techniques have been carefully analyzed in this chapter.

COMMERCIAL TECHNIQUES

Glassblowing

Glassblowing is a glassforming technique that involves inflating molten glass into a bubble (or parison) with the aid of a blowpipe (or blow tube). A person who blows glass is called a glassblower, glassmith, or gaffer. A lampworker (often also called a glassblower or glassworker) manipulates glass with the use of a torch on a smaller scale, such as in producing precision laboratory glassware out of borosilicate glass.

Technology

Principles

As a novel glass forming technique created in the middle of the 1st century BC, glassblowing exploited a working property of glass that was previously unknown to glassworkers; inflation, which is the expansion of a molten blob of glass by introducing a small amount of air to it. That is based on the liquid structure of glass where the atoms are held together by strong chemical bonds in a disordered and random network, therefore molten glass is viscous enough to be blown and gradually hardens as it loses heat.

To increase the stiffness of the molten glass, which in turn facilitates the process of blowing, there was a subtle change in the composition of glass. With reference to their studies of the ancient glass assemblages from Sepphoris of Israel, Fischer and McCray postulated that the concentration of natron, which acts as flux in glass, is slightly lower in blown vessels than those manufactured by casting. Lower concentration of natron would have allowed the glass to be stiffer for blowing.

During blowing, thinner layers of glass cool faster than thicker ones and become more viscous than the thicker layers. That allows production of blown glass with uniform thickness instead of causing blow-through of the thinned layers.

A full range of glassblowing techniques was developed within decades of its invention. The two major methods of glassblowing are free-blowing and mold-blowing.

Free-blowing

This method held a pre-eminent position in glassforming ever since its introduction in the middle of the 1st century BC until the late 19th century, and is still widely used nowadays as a glassforming technique, especially for artistic purposes. The process of free-blowing involves the blowing of short puffs of air into a molten portion of glass called a "gather" which has been spooled at one end of the blowpipe. This has the effect of forming an elastic skin on the interior of the glass blob that matches the exterior skin caused by the removal of heat from the furnace. The glassworker can then quickly inflate the molten glass to a coherent blob and work it into a desired shape.

Glassblowing demo at North California Renaissance Fair .

Researchers at the Toledo Museum of Art attempted to reconstruct the ancient free-blowing technique by using clay blowpipes. The result proved that short clay blow-pipes of about 30–60 cm (12–24 in) facilitate free-blowing because they are simple to handle and to manipulate and can be re-used several times. Skilled workers are capable of shaping almost any vessel forms by rotating the pipe, swinging it and controlling the temperature of the piece while they blow. They can produce a great variety of glass objects, ranging from drinking cups to window glass.

An outstanding example of the free-blowing technique is the Portland Vase, which is a cameo manufactured during the Roman period. An experiment was carried out by Gudenrath and Whitehouse with the aim of re-creating the Portland Vase. A full amount of blue glass required for the body of the vase was gathered on the end of the blowpipe and was subsequently dipped into a pot of hot white glass. Inflation occurred when the glassworker blew the molten glass into a sphere which was then stretched or elongated into a vase with a layer of white glass overlying the blue body.

Mold-blowing

Glassblower Jean-Pierre Canlis sculpting a
section of his piece "Insignificance".

Mold-blowing was an alternative glassblowing method that came after the invention of free-blowing, during the first part of the second quarter of the 1st century AD. A glob of molten glass is placed on the end of the blowpipe, and is then inflated into a wooden or metal carved mold. In that way, the shape and the texture of the bubble of glass is determined by the design on the interior of the mold rather than the skill of the glassworker.

Two types of molds, namely single-piece mold and multi-piece mold, are frequently used to produce mold-blown vessels. The former allows the finished glass object to be removed in one movement by pulling it upwards from the single-piece mold and is largely employed to produce tableware and utilitarian vessels for storage and transportation. Whereas the latter is made in multi-paneled mold segments that join together, thus permitting the development of more sophisticated surface modeling, texture and design.

The Roman leaf beaker which is now on display in the J. Paul Getty Museum was blown in a three-part mold decorated with the foliage relief frieze of four vertical plants. Meanwhile, Taylor and Hill tried to reproduce mold-blown vessels by using three-part molds made of different materials. The result suggested that metal molds, in particular bronze, are more effective in producing high-relief design on glass than plaster or wooden molds.

The development of the mold-blowing technique has enabled the speedy production of glass objects in large quantity, thus encouraging the mass production and widespread distribution of glass objects.

Modern Glassblowing

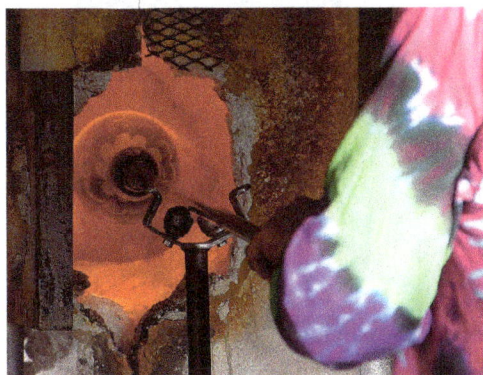

Use of a glory hole to reheat a piece on the end of a blowpipe.

The transformation of raw materials into glass takes place around 1,320 °C (2,400 °F); the glass emits enough heat to appear almost white hot. The glass is then left to "fine out" (allowing the bubbles to rise out of the mass), and then the working temperature is reduced in the furnace to around 1,090 °C (2,000 °F). At this stage, the glass appears to be a bright orange color. Though most glassblowing is done between 870 and 1,040 °C (1,600 and 1,900 °F), "soda-lime" glass remains somewhat plastic and workable as low as 730 °C (1,350 °F). Annealing is usually done between 371 and 482 °C (700 and 900 °F).

Glassblowing involves three furnaces. The first, which contains a crucible of molten glass, is simply referred to as the furnace. The second is called the glory hole, and is used to reheat a piece in between steps of working with it. The final furnace is called the lehr or annealer, and is used to slowly cool the glass, over a period of a few hours to a few days, depending on the size of the pieces. This keeps the glass from cracking or shattering due to thermal stress. Historically, all three furnaces were contained in one structure, with a set of progressively cooler chambers for each of the three purposes.

The major tools used by a glassblower are the blowpipe (or blow tube), punty (or punty rod, pontil, or mandrel), bench, marver, blocks, jacks, paddles, tweezers, newspaper pads, and a variety of shears.

The tip of the blowpipe is first preheated; then dipped in the molten glass in the furnace. The molten glass is "gathered" onto the end of the blowpipe in much the same way that viscous honey is picked up on a honey dipper. This glass is then rolled on the marver, which was traditionally a flat slab of marble, but today is more commonly a fairly thick flat sheet of steel. This process, called marvering, forms a cool skin on the exterior of the molten glass blob, and shapes it. Then air is blown into the pipe, creating a bubble. Next, the glassworker can gather more glass over that bubble to create a larger piece. Once a piece has been blown to its approximate final size, the bottom is finalized. Then, the molten glass is attached to a stainless steel or iron rod called a punty for shaping and transferring the hollow piece from the blowpipe to provide an opening to finalize the top.

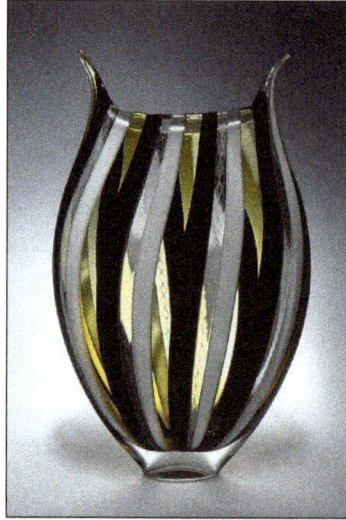

Glass can be made with precise striped patterns through a process
called cane which involves the use of rods of colored glass.

The bench is a glassblower's workstation, and has a place for the glassblower to sit, a place for the handheld tools, and two rails that the pipe or punty rides on while the blower works with the piece. Blocks are ladle-like tools made from water-soaked fruit-wood, and are used similarly to the marver to shape and cool a piece in the early steps of creation. In similar fashion, pads of water-soaked newspaper (roughly 15 cm (6 in) square, 1.3 to 2.5 centimetres (0.5 to 1 in) thick), held in the bare hand, can be used to shape the piece. Jacks are tools shaped somewhat like large tweezers with two blades, which are used for forming shape later in the creation of a piece. Paddles are flat pieces of wood or graphite used for creating flat spots such as a bottom. Tweezers are used to pick out details or to pull on the glass. There are two important types of shears, straight shears and diamond shears. Straight shears are essentially bulky scissors, used for making linear cuts. Diamond shears have blades that form a diamond shape when partially open. These are used for cutting off masses of glass.

There are many ways to apply patterns and color to blown glass, including rolling molten glass in powdered color or larger pieces of colored glass called frit. Complex patterns with great detail can be created through the use of cane (rods of colored glass) and murrine (rods cut in cross-sections to reveal patterns). These pieces of color can be arranged in a pattern on a flat surface, and then "picked up" by rolling a bubble of molten glass over them. One of the most exacting and complicated caneworking techniques is "reticello", which involves creating two bubbles from cane, each twisted in a different direction and then combining them and blowing out the final form.

A lampworker, usually operating on a much smaller scale, historically used alcohol lamps and breath or bellows-driven air to create a hot flame at a workbench to manipulate preformed glass rods and tubes. These stock materials took form as laboratory glassware, beads, and durable scientific "specimens"—miniature glass sculpture. The craft, which was raised to an art form in the late 1960s by Hans Godo Frabel (later followed by lampwork artists such as Milon Townsend and Robert Mickelson), is still practiced today. The modern lampworker uses a flame of oxygen and propane or natural gas. The modern torch permits working both the soft glass from the furnace worker and the borosilicate glass (low-expansion) of the scientific glassblower. This latter worker may also have multiple headed torches and special lathes to help form the glass or fused quartz used for special projects.

Precision Glass Moulding

Precision glass moulding is a replicative process that allows the production of high precision optical components from glass without grinding and polishing. The process is also known as ultra-precision glass pressing. It is used to manufacture precision glass lenses for consumer products such as digital cameras, and high-end products like medical systems. The main advantage over mechanical lens production is that complex lens geometries such as aspheres can be produced cost-efficiently.

Process

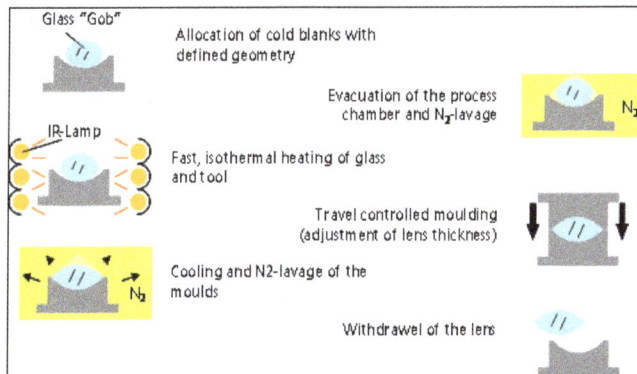

Summary of process.

The precision glass moulding process consists of six steps:

- The glass blank is loaded into the lower side of the moulding tool.

- Oxygen is removed from the working area by filling with nitrogen and evacuation of the process chamber.

- The tool system is nearly closed (no contact of the upper mould) and the entire system of mould, die and glass is heated up. Infrared lamps are used for heating in most systems.

- After reaching the working temperature, which is between the transition temperature and the softening point of the glass, the moulds close further and start pressing the glass in a travel-controlled process.

- When the final thickness of the part has been achieved, the pressing switches over to a force-controlled process.

- After moulding has been completed, the glass is cooled down and the working environment is filled with nitrogen. When the lens has cooled to the point where it can be handled, it is removed from the tool.

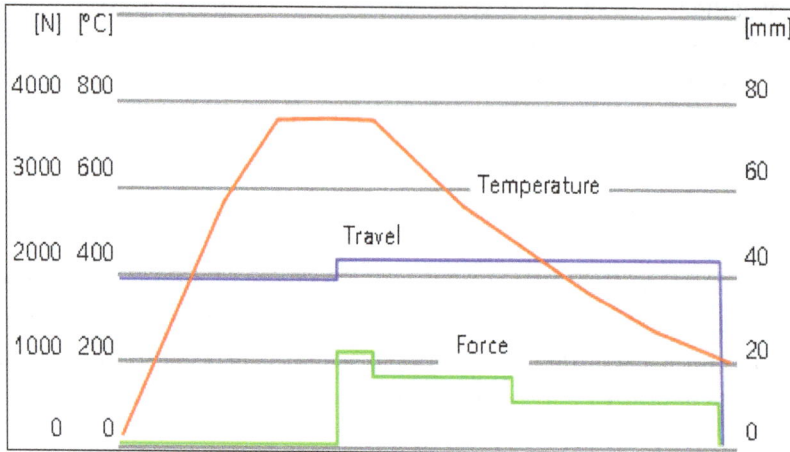

Temperature (in °C), travel (in mm), and force (in N) during the process.

The process is executed on a specialized moulding machine, which precisely controls the temperature, travel, and force during the process. The tools used must withstand high temperatures and pressures, and need to be resistant to chemical interaction with the glass. The mold materials also have to be suitable for machining into the precise surface profiles.

Process Chain

In order to ensure the desired quality the parts are measured between each process step. Additionally, the parts are handled and transported carefully between the processing and metrology steps.

- Hotforming of gobs: The precision glass moulding process yields the best results in both quality and cost if it works with precise preforms. These are usually acquired by pressing or hotforming of "gobs" of molten glass. This step is done by continuous glass melting and moulding in single-sided metal moulds. This process is only suitable for high production volumes. For smaller production volumes, the preforms have to be manufactured by mechanical material-removing steps from blocks or slices of raw glass.

- Precision glass moulding: In this step the preform is directly formed into an optical glass lens. It is necessary to clean the glass preform and the mould before starting the process, but there is no polishing or post-machining required.

- Lens coating: An antireflection coating is applied to the finished lenses. The lenses are first cleaned, and then loaded into a fixture. The fixture, containing a large number of lenses, is placed in the coating machine. After finishing the process the glass lenses are removed from the holder and the holder is cleaned by sand-blasting or other techniques. Usually the optical coating is done by one of two methods: physical vapour deposition (PVD), in which oxide materials evaporate and are deposited on the lens, and plasma-enhanced chemical vapor deposition (PECVD). Chemical reactions take place in a vacuum and the reaction product is deposited on the lens. The lenses are coated for two reasons:

- ∘ Manipulate or improve the optical transmission / reflection.

- ∘ Enhance the mechanical, chemical or electrical properties.

Tool and Mould Design

Shape of Optical Element

Precision glass moulding can be used to produce a large variety of optical form elements such as spheres, aspheres, free-form elements and array-structures.

Concerning the curvature of the lens elements, the following statements can be drawn: Acceptable lens shapes are most bi-convex, plano-convex and mild meniscus shapes. Not unacceptable but hard to mould are bi-concave lenses, steep meniscus lenses, and lenses with severe features (e.g. a bump on a convex surface). In general, plano-curved lenses are easier to mould than lenses with both sides curved since matching of flat faces is easier. Moulding concave forms with small centre thickness is difficult due to sticking of the moulded part to the mould occurring as a result of the different thermal expansion coefficients. Furthermore it is recommended to avoid undercuts and sharp edges. For the lens design it should be considered that the lens has to be mountable in measurement systems.

Shape of Preforms

The shape of the preform or "blank" needs to be chosen according to the geometry of the finished optical element. Possible preforms are spherical (ball), near spherical (gob), plano-plano, plano-convex, plano-concave, biconvex and biconcave blanks. Ball and gob-blanks do not have to be premachined whereas other preforms require grinding and polishing.

The following topic describes basic traits of preform choice:

- Formed Ball Preform: "Used specifically for lenses with positive power: biconvex, plano-convex, and meniscus where the convex side is stronger than the concave side, this only works for a relatively small volume of material."

- Ground and Polished Plano-Plano Preform: "As a lens changes to negative in power biconcave, plano-concave, and meniscus where the concave side is stronger, an alternative preform shape, plano-plano, is required for the molding process. Relative to a formed preform an increase in cost is observed for the manufacturing of this type of preform."

- Ground and Polished Ball Preform: "When the geometry of a lens extends beyond the volume range of a formed ball preform, a ground and polished ball preform is required. Used for lenses with positive power: biconvex, plano-convex, and meniscus: where the convex side is stronger, this geometry allows for molding of lenses with larger total volume. Relative to a formed preform and a plano-plano preform, an increase in cost is observed for the manufacturing of this type of preform."

- Ground and Polished Lenslet Preform: "The Lenslet preform is primarily for lenses with positive power, biconvex, planoconvex, and meniscus: where the convex side is the strongest surface. The use of this type of preform allows for molding of the largest volume of glass at any given time in the molding machines. The Lenslet is traditionally ground and polished to a near net shape of the final lens, and then pressed. The cost associated with the manufacturing of the lenslet preform is the highest of all preform types."

- Gob Preform: Precision gobs can be used as preforms for the production of aspherical lenses in a precision molding process. They are manufactured from a continuous glass melting process. The resulting precision gobs exhibit a very smooth firepolished surface with an excellent surface roughness and high volume accuracy.

Dimensions

The dimensions of the optical elements that can be moulded depend on the size of the moulding machine. The precision glass moulding process is not limited to small optics. For the right element geometry, it can enable economical production of aspheric lenses up to 60 mm in diameter and more than 20 mm thick.

General design recommendations:

Size:

- Diameter range: 0.5–70 mm depending on the application.

- Flank angle: Sometimes <60 degree due to limited metrology but higher angles are possible by some manufacturers that have expanded metrology capabilities (e.g. Panasonic UA3P or similar).

- Edge thickness preferably > 1.0 mm, alternatively 0.5 to 2.0 x center thickness

- Clear aperture should be smaller than the lens diameter, preferably at least 1.0 mm (per side) less.

- Optical surfaces.

Radius:

- Base radius no less than 3.0 mm.

Optical Surfaces:

- Sags no greater than 8 mm on both concave and convex surfaces.

- Transition from the optical surface to the lens outside diameter requires a minimum radius value of 0.3 mm.

Volume:

- Volume of the lens (including flanges), $V \leq 4/3 \, \pi \, r^3$, where r is the smallest local convex radius.

Tolerances

Although the form, dimensional and positional tolerances that can be achieved in precision glass moulding are subject to a natural border, the values being achieved in practice strongly depend on the degree of control and experience in mould making and moulding. The table below gives an overview of achievable manufacturing tolerances in precision glass moulding at different companies.

Parameter	Rochester Precision Optics	Braunecker	Lightpath Optics	Ingeneric	FISBA
Diameter	+0/-0.010 mm		+/-0.005 mm	0,005 mm	+/- 0.005 mm
Center thickness	+/-0,012 mm		+/- 0.010 mm	0,010 mm	+/- 0.01 mm
Alignment	2,5 min (axis)	< 1.5min (angular)	-	-	5 µm
Scratch-Dig	20-10		20-10/10-5	-	20-10
Abbe-Number	+/-0.5%		+/-1%	1%	-
Surface Roughness	-	<3 nm	10 nm	5 nm	4 nm
Index of Refraction	+/-0.0003		+/-0.001	0,001	-
Wedge	0.01mm		+/-1 arcmin	1 arcmin	3'

For aspherical lenses, the design should be able to tolerate 0.010 mm of lateral shear between surfaces plus 5 micrometres Total Internal Reflection of wedge (across the part without considering the lateral shear) to be considered robust.

Specifications for aspheres:

- Surface roughness (Ra): < 3 µm depending on diameter.

- Form error (PV): < 1 µm depending on diameter.

Index Drop

Due to the fast cooling after moulding, the part retains a small amount of residual stress. Consequently, the glass exhibits a small change in the refractive index which has to be considered in the optical design. A higher cooling rate corresponds to a larger decrease of the refractive index. A lower cooling rate could circumvent the index drop, but would be less cost-efficient.

Glass Material

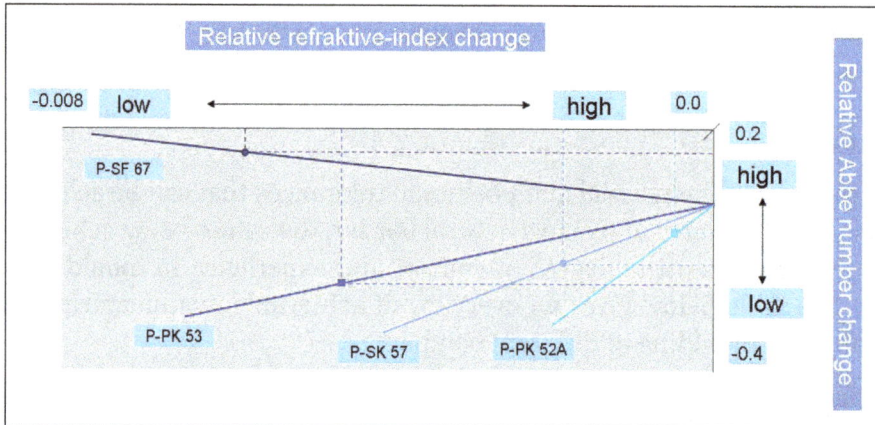

Change in refractive index and Abbe number for different glass types and annealing rates.

Many glasses can be used with PGM. However, there are some limitations:

- The glass transition temperature Tg must not exceed the maximum heating temperature of the mould.

- Many lead oxide flint glasses are not compliant with the RoHS EU directive (*Restriction of certain Hazardous Substances*).

- The glass composition influences moulding tool life.

- Chalcogenide materials require certain preform shapes.

- Glass expansion/contraction is highly temperature and rate dependent phenomenon. The coefficient of thermal expansion (CTE) of mould and glass should match. High CTE difference means high deviation between the moulded glass and the mould. High CTE glasses are also critical in terms of non-uniform temperature distribution in the glass. This means that especially fast cooling can not be applied. In addition to this, the temperature difference between the warm lens directly after moulding and the surrounding air can easily cause cracks.

- In addition, temperature dependence of viscosity of glass, structural and stress viscoelastic relaxation of glass play an important role in determination of lens preform shape, final state of stress and shape deviation.

- The internal and external quality of the blank must be the same or better than the requirements of the finished lens since the precision glass moulding process is not able to improve the glass quality.

- The glass exhibits a change in refractive index, called index drop, during the annealing process. This drop is caused by fast cooling of the mould insert, inducing a small amount of residual stress in the glass. As a result, the glass exhibits a small index change when compared to its fine anneal state. The index drop is small (usually .002-.006), but the optical design needs to be optimized to compensate for this change. As an example, the index drop for different glass types is displayed in the picture on the right for different annealing rates. Note that the annealing rate is not necessarily constant during the cooling process. Typical "average" annealing rates for precision molding are between 1000 K/h and 10,000 K/h (or higher). Not only the refractive index, but also the Abbe-number of the glass is changed due to fast annealing.

So-called "low-Tg-glasses" with a maximum transition temperature of less than 550 °C have been developed in order to enable new manufacturing routes for the moulds. Mould materials such as steel can be used for moulding low-Tg-glasses whereas high-Tg-glasses require a high-temperature mould material, such as tungsten carbide.

Substrate Materials

The mould material must have sufficient strength, hardness and accuracy at high temperature and pressure. Good oxidation resistance, low thermal expansion and high thermal conductivity are also required. The material of the mould has to be suitable to withstand the process temperatures without undergoing deforming processes. Therefore, the mould material choice depends critically on the transition temperature of the glass material. For low-Tg-glasses, steel moulds with a nickel alloy coating can be used. Since they cannot withstand the high temperatures required for regular optical glasses, heat-resistant materials such as carbide alloys have to be used instead in this case. In addition, mould materials include aluminium alloys, glasslike or vitreous carbon, silicon carbide, silicon nitride and a mixture of silicon carbide and carbon.

A commonly used material in mould making is tungsten carbide. The mould inserts are produced by means of powder metallurgy, i.e. a sintering process followed by post-machining processes and sophisticated grinding operations. Most commonly a metallic binder (usually cobalt) is added in liquid phase sintering. In this process, the metallic binder improves the toughness of the mould as well as the sintering quality in the liquid phase to fully dense material. Moulds made of hard materials have a typical lifetime of thousands of parts (size dependent) and are cost-effective for volumes of 200-1000+ (depending upon the size of the part).

Mould Manufacturing

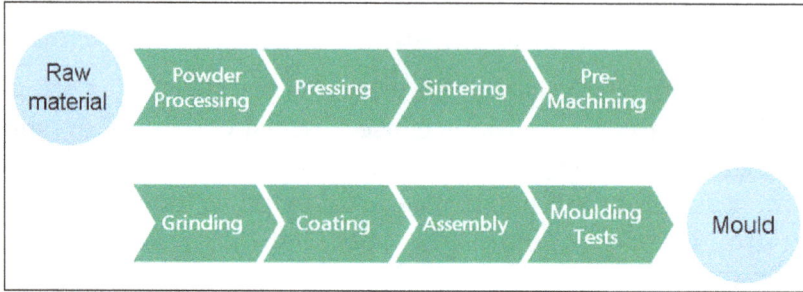

In order to ensure high quality standards metrology steps are implemented between each process step.

1. Powder processing: This process step is responsible for achieving grain sizes suitable for pressing and machining. The powder is processed by milling the raw material.

2. Pressing: This step does the pre-forming of "green" raw bodies of the mould inserts.

3. Sintering: By sintering, the pre-formed green bodies are compressed and hardened. In order to do this the green body is heated to a temperature below the melting temperature. The sintering process consists of three phases: First, the volume and the porosity is reduced and secondly, the open porosity is reduced. In the third phase, sinter necks are formed which enhance the material's strength.

4. Pre-Machining: The step of Pre-Machining creates the main form of the optical insert. It typically contains four process steps. These steps are grinding the inner/outer diameter, grinding the parallel/end faces of the insert, grinding/lapping of the fitting of insert, and finally, the near-net-shape grinding of the cavity. Normally, the cavity is only pre-machined to a flat or a best-fit sphere.

5. Grinding: Grinding or finish-machining creates the final form and the surface finish of the cavity in the mould insert. Usually, the finish is carried out by grinding; a subsequent polishing step is optionally required. Finish grinding can require several changes of the grinding tool and several truing steps of the tool. Finish-machining of the mould is an iterative process: As long as the machined mould shows deviations from the nominal contour in the measurement step after grinding, it has to be reground. There is no well-defined border between pre-machining and fine grinding. Throughout the grinding process of the cavity, the grain size of the tool, the feed rate and the cutting depth are reduced whereas machining time increases. Convex surfaces are easier to manufacture. The necessary steps of workpiece preparation are the mould alignment and the mould referencing. Grinding tool alignment, grinding tool referencing and grinding tool truing also have to be done. After that polishing can be necessary

to remove the anisotropic structure which remains after grinding. It can be performed manually or by a CNC-machine.

6. Coating: Coating is the process step in which a layer is applied on the cavity surface of the optical insert which protects the mould against wear, corrosion, friction, sticking of glass and chemical reactions with glass. For coating the surface of moulds by physical vapour deposition (PVD), metals are evaporated in combination with process-gas-based chemicals. On the tool surface, highly adherent thin coatings are synthesized. Materials for coatings on optical inserts are Platinum-based PVD (mostly iridium-alloyed, standard), diamond-like carbon (not yet commercially available), SiC (CVD) on SiC-ceramics (not yet commercially available, have to be post-machined) or TiAlN (not yet commercially available). To achieve a homogeneous layer thickness, the mould's position has to be changed during coating. To prepare the mould for the coating the surfaces have to be degreased, cleaned (under clean room or near-clean room conditions) and batched. Especially the cathode of the machine has to be cleaned. After this process the workpiece has to be debatched.

7. Assembly: In this process step the optical insert and the mould base are combined to the assembled mould. For one optical element two mould inserts are necessary which are assembled outside the machine. For the assembly height measurement and spacer adjustment are essential.

8. Moulding Tests: This step determines whether the mould creates the specified form and surface quality. If mould is not suitable, it has to be reground. It is part of an iterative loop. The assembly of the mould has to be put into the machine to start the try-out-moulding.

In order to save the quality and enable an early warning in case of any problems between every single step there has to be a step of measurement and referencing. Besides that the time for transport and handling has to be taken into account in the planning of the process.

Metrology and Quality Assurance

Once process and tool have been developed, precision glass moulding has a great advantage over conventional production techniques. The majority of the lens quality characteristics are tool-bound. This means that lenses, which are pressed with the same tool and process, usually have only insignificantly small deviations. For example, an important characteristic of a lens is the form of the optical surface. In the case of aspherical lenses the measurement of optical surfaces is very difficult and connected to high efforts. Additionally, when working with tactile measurement systems there is always a risk that the optical surface might be scratched. For precision moulded lenses such measurements are only necessary for a small amount of sample lenses in order to qualify the tool. The series production can then be executed without further need for measurements. In this case, only the cleanliness of the optical surface has to be monitored. Another advantage is that the lens' center thickness can be estimated from the easily measurable edge thickness or by applying a contactless measurement system.

Protective Coatings

In order to enhance the mould insert's lifetime, protective coatings can be applied. "The materials that have been selected for the antistick coatings can be divided into 5 groups including: (1) single layer carbides, nitrides, oxides and borides such as TiN, BN, TiAlN, NiAlN, TiBC, TiBCN, NiCrSiB and Al_2O_3, (2) nitrides or carbides based gradient and multilayers, (3) nitrides based superlattice films, (4) amorphous carbon or diamond-like carbon and (5) precious metal based alloys."

Experiments carried out by Ma et al. yield the following results: "The higher the temperature, the smaller the wetting angle between glass gob and substrate could be observed. This indicates that severe interface chemical reaction occurred and resulted in the loss of transparency in glass appearance. The wetting experiment in nitrogen ambient improved the sticking situation. The combination of chemically stable substrates and coatings, such as Sapphire (substrate) / GaN (film) and Glass (substrate) / Al_2O_3 (film) can achieve the best antistick propose. The precious metal films such as PtIr(Platinum, Iridium) coated on the ceramic substrates can effectively reduce the interface reaction between the glass and substrates."

Glass Casting

Glass casting is the process in which glass objects are cast by directing molten glass into a mould where it solidifies. The technique has been used since the Egyptian period. Modern cast glass is formed by a variety of processes such as kiln casting, or casting into sand, graphite or metal moulds.

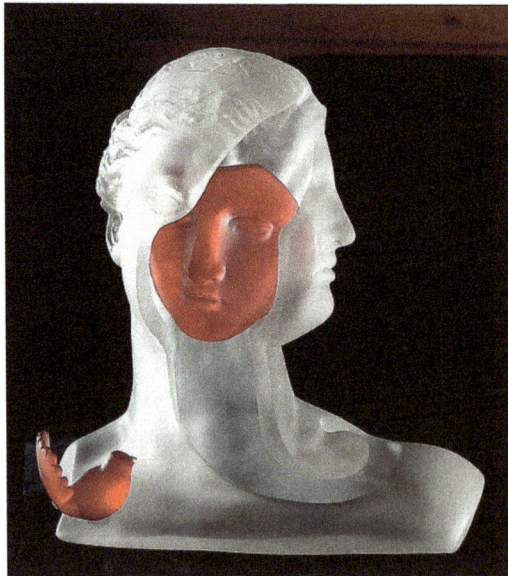

A cast glass sculpture from a kiln firing.

Modern Techniques

Sand Casting

Sand casting involves the use of hot molten glass poured directly into a preformed mould. It is a process similar to casting metal into a mould. The sand mould is typically prepared by using a mixture of clean sand and a small proportion of the water-absorbing clay bentonite. Bentonite acts as a binding material. In the process, a small amount of water is added to the sand-bentonite mixture and this is well mixed and sifted before addition to an open topped container. A template is prepared (typically made of wood, or a found object or even a body part such as a hand or fist) which is tightly pressed into the sand to make a clean impression. This impression then forms the mould.

The surface of the mould can be covered in coloured glass powders or frits to give a surface colour to the sand cast glass object. When the mould preparation is complete hot glass is ladled from the furnace at temperatures of about 1,200 °C (2,190 °F) to allow it to freely pour. The hot glass is poured directly into the mould. During the pouring process, glass or compatible objects may be placed to later give the appearance of floating in the solid glass object. This very immediate and dynamic method was pioneered and perfected in the 1980s by the Swedish artist Bertil Vallien.

Kiln Casting

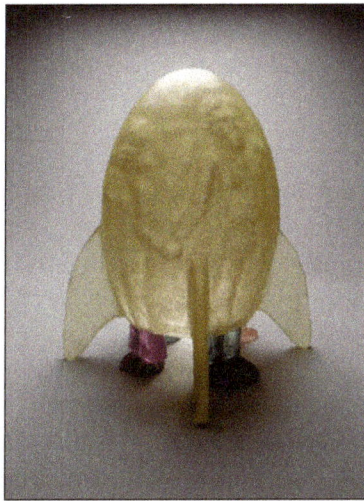

Kiln-Cast lead crystal 'Nuclear Family'.

Kiln casting involves the preparation of a mould which is often made of a mixture of plaster and refractory materials such as silica. A model can be made from any solid material, such as wax, wood, or metal, and after taking a cast of the model (a process called investment) the model is removed from the mould. One method of forming a mould is by the *Cire perdue* or "lost wax" method. Using this method, a model can be made from wax and after investment the wax can be steamed or burned away in a kiln, forming a cavity. The heat resistant mould is then placed in a kiln equipped with a funnel-like opening filled with solid glass granules or lumps. The kiln is heated to between 800 °C (1,470 °F) and 1,000 °C (1,830 °F), and as the glass melts it runs into and fills the mould.

Pâte de Verre

Pâte de verre is a form of kiln casting and literally translated means glass paste. In this process, finely crushed glass is mixed with a binding material, such as a mixture of gum arabic and water, and often with colourants and enamels. The resultant paste is applied to the inner surface of a negative mould forming a coating. After the coated mould is fired at the appropriate temperature the glass is fused creating a hollow object that can have thick or thin walls depending on the thickness of the pate de verre layers. Daum, a French commercial crystal manufacturer, produce highly sculptural pieces in pate de verre.

Three pate de verre vessels.

Graphite Casting

Graphite is also used in the hot forming of glass. Graphite moulds are prepared by carving into them, machining them into curved forms, or stacking them into shapes. Molten glass is poured into a mould where it is cooled until hard enough to be removed and placed into an annealing kiln to cool slowly.

Vitrification

Vitrification is a process of converting a material into a glass-like amorphous solid that is free from any crystalline structure, either by the quick removal or addition of heat, or by mixing with an additive. Solidification of a vitreous solid occurs at the glass transition temperature (which is lower than melting temperature, T_m, due to supercooling).

When the starting material is solid, vitrification usually involves heating the substances to very high temperatures. Many ceramics are produced in such a manner. Vitrification may also occur naturally when lightning strikes sand, where the extreme and immediate heat can create hollow, branching rootlike structures of glass, called fulgurite. When applied to whiteware ceramics, vitreous means the material has an extremely low permeability to liquids, often but not always water, when determined by a specified test regime. The microstructure of whiteware ceramics frequently contain both amorphous and crystalline phases.

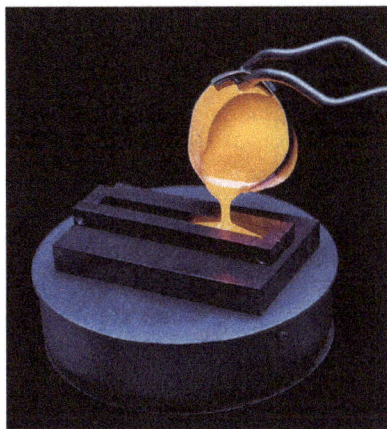

A vitrification experiment for the study of nuclear waste
disposal at Pacific Northwest National Labs.

Examples:

When sucrose is cooled slowly, the result is crystal sugar (or rock candy), but, when cooled rapidly, the result can be in the form of syrupy cotton candy (candyfloss). Vitrification can also occur when starting with a liquid such as water, usually through very rapid cooling or the introduction of agents that suppress the formation of ice crystals. Additives used in cryobiology or produced naturally by organisms living in polar regions are called cryoprotectants. Arctic frogs and some other ectotherms naturally produce glycerol or glucose in their livers to reduce ice formation. When glucose is used as a cryoprotectant by Arctic frogs, massive amounts of glucose are released at low temperature, and a special form of insulin allows for this extra glucose to enter the cells. When the frog rewarms during spring, the extra glucose must be rapidly removed from the cells and recycled via renal excretion and storage in the bladder. Arctic insects also use sugars as cryoprotectants. Arctic fish use antifreeze proteins, sometimes appended with sugars, as cryoprotectants.

Applications

Ordinary soda-lime glass, used in windows and tableware, is created by the addition of sodium carbonate and lime (calcium oxide) to silicon dioxide. Without these additives, silicon dioxide will (with slow cooling) form sand or quartz crystal, not glass.

Vitrification is a proven technique in the disposal and long-term storage of nuclear waste or other hazardous wastes. Waste is mixed with glass-forming chemicals to form molten glass that then solidifies, immobilizing the waste. The final waste form resembles obsidian and is a non-leaching, durable material that effectively traps the waste inside. The waste can be stored for relatively long periods in this form without concern for air or groundwater contamination. Bulk vitrification uses electrodes to melt soil and wastes where they lie buried. The hardened waste may then be disinterred with less danger of widespread contamination. According to the Pacific Northwest National Labs, "Vitrification locks dangerous materials into a stable glass form that will last for thousands of years."

Ethylene glycol is used as automotive antifreeze and propylene glycol has been used to reduce ice crystals in ice cream, making it smoother.

For years, glycerol has been used in cryobiology as a cryoprotectant for blood cells and bull sperm, allowing storage at liquid nitrogen temperatures. However, glycerol cannot be used to protect whole organs from damage. Instead, many biotechnology companies are currently researching the development of other cryoprotectants more suitable for such uses. A successful discovery may eventually make possible the bulk cryogenic storage (or "banking") of transplantable human and xenobiotic organs. A substantial step in that direction has already occurred. At the July 2005 annual conference of the Society for Cryobiology, Twenty-First Century Medicine announced the vitrification of a rabbit kidney to -135°C with their proprietary vitrification cocktail. Upon rewarming, the kidney was successfully transplanted into a rabbit, with complete functionality and viability.

In the context of cryonics, especially in preservation of the human brain, vitrification of tissue is thought to be necessary to prevent destruction of the tissue or information encoded in the brain. At present, vitrification techniques have only been applied to brains (neurovitrification) by Alcor and to the upper body by the Cryonics Institute, but research is in progress by both organizations to apply vitrification to the whole body.

ARTISTIC TECHNIQUES

Caneworking

In glassblowing, cane refers to rods of glass with color; these rods can be simple, containing a single color, or they can be complex and contain strands of one or several colors in pattern. Caneworking refers to the process of making cane, and also to the use of pieces of cane, lengthwise, in the blowing process to add intricate, often spiral, patterns and stripes to vessels or other blown glass objects. Cane is also used to make murrine (singular "murrina", sometimes called mosaic glass), thin discs cut from the cane in cross-section that are also added to blown or hot-worked objects. A particular form of murrine glasswork is millefiori ("thousand flowers"), in which many murrine with a flower-like or star-shaped cross-section are included in a blown glass piece.

Caneworking is an ancient technique, first invented in southern Italy in the second half of the third century BC, and elaborately developed centuries later on the Italian island of Murano.

Making Cane

There are several different methods of making cane. In each, the fundamental technique is the same: a lump of glass, often containing some pattern of colored and clear glass, is heated in a furnace (glory hole) and then pulled, by means of a long metal rod (punty) attached at each end. As the glass is stretched out, it retains whatever cross-sectional pattern was in the original lump, but narrows quite uniformly along its length (due to the skill of the glassblowers doing the pulling, aided by the fact that if the glass becomes narrower at some point along the length, it cools more there and thus becomes stiffer). Cane is usually pulled until it reaches roughly the diameter of a pencil,when, depending on the size of the original lump, it may be anywhere from one to fifty feet in length. After cooling, it is broken into sections usually from four to six inches long, which can then be used in making more complex canes or in other glass-blowing techniques.

The simplest cane, called *vetro a fili* (glass with threads) is clear glass with one or more threads of colored (often white) glass running its length. It is commonly made by heating and shaping a chunk of clear, white, or colored glass on the end of a punty, and then "gathering" molten clear glass over the color by dipping the punty in a furnace containing the clear glass. After the desired amount of clear glass is surrounding the color, this cylinder of hot glass is then shaped, cooled and heated until uniform in shape and temperature. Simultaneously an assistant prepares a 'post' which is another punty with a small platform of clear glass on the end. The post is pressed against the end of the hot cylinder of glass to connect them, and the glassblower (or 'gaffer') and assistant walk away from each other with the punties, until the cane is stretched to the desired length and diameter. The cane cools within minutes and is cut into small sections.

Variations in Cane Making

Close-up of ballotini cane forming a part of a blown vessel.

A simple single-thread cane can then be used to make more complex canes. A small bundle of single-thread canes can be heated until they fuse, or heated canes, laid parallel, can be picked up on the circumference of a hot cylinder of clear or colored glass. This bundle, treated just as the chunk of color in the description above, is cased in clear glass and pulled out, forming a *vetro a fili* cane with multiple threads and perhaps a clear or solid color core. If the cane is twisted as it is pulled, the threads take a spiral shape called *vetro a retorti* (twisted glass) or *zanfirico*.

Ballotini is a cane technique in which several *vetro a fili* canes are picked up while laid side-by-side rather than a bundle, with a clear glass gather over them. This gather is shaped into a cylinder with the canes directed along the axis, so that the canes form a sort of "fence" across the diameter of the cylinder. When this is simultaneously twisted and pulled, the resulting cane has a helix of threads across its thickness.

A small (1 ½") disc of millefiori-patterned glass. Each of
the stars and flowers is a cross-section of a cane.

Another technique for forming cane is to use optic molds, to make more complex cross sections. An optic mold is an open-ended cone-shaped mold with some sort of lobed or star shape around its inside circumference. When a gather or partially blown bubble is forced into the mold, its outside takes the shape of the mold. Canes with complicated, multi-colored patterns are formed by placing layers of different or alternating colors over a solid-color core, using various optic molds on the layers as they are built. Because the outer layers are hotter than those inside when the molds are used, the mold shape is impressed into the outer color without deforming the inner shapes. Canes made in this way are used in making *millefiori*. Discs from eight different canes have been used to make the pendant in the photo.

Finally, flameworkers sometimes make cane by building up the cross-section using ordinary flameworking or bead making techniques. This permits very subtle gradations of color and shading, and is the way murrine portraits are usually made.

Cane Use

The generic term for blown glass made using canes in the lengthwise direction is *filigrano* (filigree glass), as contrasted with *murrine* when the canes are sliced and used in cross-section. (An older term is *latticino*, which has fallen into disuse).

One way glassblowers incorporate cane into their work is to line up canes on a steel or ceramic plate and heat them slowly to avoid cracking. When the surfaces of the canes just begin to melt, the canes adhere to each other. The tip of a glassblowing pipe (blowpipe) is covered with a 'collar' of clear molten glass, and touched to one corner of the aligned canes. The tip of the blowpipe is then rolled along the bottom of the canes, which stick to the collar, aligned cylindrically around the edge of the blowpipe. They are heated further until soft enough to shape. The cylinder of canes is sealed at the bottom with jacks and tweezers, to form the beginning of a bubble. The bubble is then blown using traditional glassblowing techniques.

Cane can also be incorporated in larger blown glass work by picking it up on a bubble of molten clear glass. This technique involves the gaffer creating a bubble from molten clear glass while an assistant heats the pattern of cane. When the cane design is fused and at the correct temperature and the bubble is exactly the correct size and temperature, the bubble is rolled over the cane pattern, which sticks to the hot glass. The bubble must be the right size and temperature for the pattern to cover it fully without any gaps or trapping air. Once the canes have been picked up, the bubble can be further heated, blown, and smoothed and shaped on the marver to give whatever final shape the glassblower wishes, with an embedded lacy pattern from the canes. Twisting the object as it is being shaped imparts a spiral shape to the overall pattern.

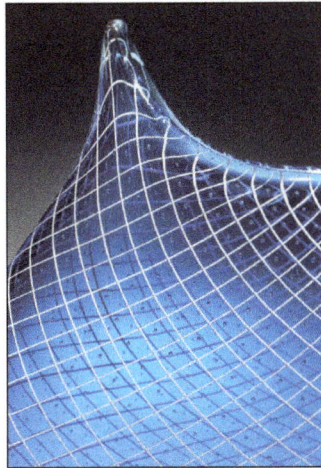

Close-up of reticello vessel blown by artist David Patchen.

The classical *reticello* pattern is a small uniform mesh of white threads in clear glass, with a tiny air bubble in every mesh rectangle. To make an object in this pattern, the glassblower first uses white single-thread *vetro a fili* canes to blow a cylindrical cup shape, twisting as he forms it so the canes are in a spiral, and using care not to totally smooth the inside ribbing that remains from the canes. Setting this cup aside (usually keeping it warm in a furnace, below its softening point), he then makes another closed cylinder in the same pattern, but twisted in the opposite direction, and retaining some of the ribbing on the cylinder's outside. When this cylinder is the right size, the glassblower plunges it into the warm cup, without touching any of the sides until it is inserted all the way. Air is trapped in the spaces between the ribs of the two pieces, forming the uniformly spaced air bubbles. The piece may then be blown out and shaped as desired. The term *reticello* is often loosely applied to any criss-cross pattern, whether *vetro a fili* or *vetro a retorti* , white or colored, and with or without air bubbles.

Traditional canework.

Contemporary canework.

Glass Beadmaking

The technology for glass beadmaking is among the oldest human arts, dating back 3,000 years. Glass beads have been dated back to at least Roman times. Perhaps the earliest glass-like beads were Egyptian faience beads, a form of clay bead with a self-forming vitreous coating. Glass beads are significant in archaeology because the presence of glass beads often indicate that there was trade and that the beadmaking technology was being spread. In addition, the composition of the glass beads could be analyzed and help archaeologists understand the sources of the beads.

Lampwork glass beads.

Common Types of Glass Bead Manufacture

Glass beads are usually categorized by the method used to manipulate the glass - wound beads, drawn beads, and molded beads. There are composites, such as millefiori beads, where cross-sections of a drawn glass cane are applied to a wound glass core. A very minor industry in blown glass beads also existed in 19th-century Venice and France.

Wound Glass Beads

Probably the earliest beads of true glass were made by the winding method. Glass at a temperature high enough to make it workable, or "ductile", is laid down or wound around a steel wire or mandrel coated in a clay slip called "bead release." The wound bead, while still hot, may be further shaped by manipulating with graphite, wood, stainless steel, brass, tungsten or marble tools and paddles. This process is called marvering, originating from the French word "marbrer" which translates to "marble". It can also be pressed into a mold in its molten state. While still hot, or after re-heating, the surface of the bead may be decorated with fine rods of colored glass called stringers creating a type of lampwork bead.

Drawn Glass Beads

The drawing of glass is also very ancient. Evidence of large-scale drawn-glass bead-making has been found by archeologists in India, at sites like Arekamedu dating to the 2nd century CE. The small drawn beads made by that industry have been called Indo-Pacific beads, because they may have been the single most widely traded item in history—found from the islands of the Pacific to Great Zimbabwe in southern Africa.

There are several methods for making drawn beads, but they all involve pulling a strand out of a gather of glass in such a way as to incorporate a bubble in the center of the strand to serve as the hole in the bead. In Arekamedu this was accomplished by inserting a hollow metal tube into the ball of hot glass and pulling the glass strand out around it, to form a continuous glass tube. In the Venetian bead industry, molten glass was gathered on the end of a tool called a puntile ("puntying up"), a bubble was incorporated into the center of a gather of molten glass, and a second puntile was attached before stretching the gather with its internal bubble into a long cane. The pulling was a skilled process, and canes were reportedly drawn to lengths up to 200 feet (61 m) long. The drawn tube was then chopped, producing individual drawn beads from its slices. The resulting beads were cooked or rolled in hot sand to round the edges without melting the holes closed; were sieved into sizes; and, usually, strung onto hanks for sale.

The most common type of modern glass bead is the seed bead, a small type of bead typically less than 6 mm, traditionally monochrome, and manufactured in very large quantities.They are a modern example of mechanically-drawn glass beads. The micro-bead or "seed bead", are so called due to their tiny, regular size. Modern seed beads are extruded by machine and some, such as Miyuki delicas, look like small tubes.

Molded Beads

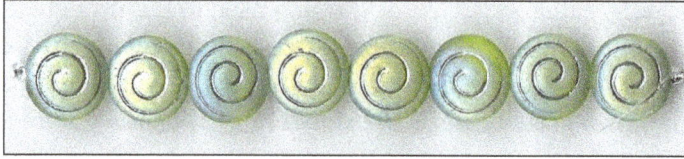

Pressed glass beads.

Pressed or molded beads are associated with lower labour costs. These are made in the Czech republic. Thick rods are heated to molten and fed into a complex apparatus that stamps the glass, including a needle that pierces a hole. The beads again are rolled in hot sand to remove flashing and soften seam lines. By making canes (the glass rods fed into the machine) striped or otherwise patterned, the resulting beads can be more elaborately colored than seed beads. One 'feed' of a hot rod might result in 10–20 beads, and a single operator can make thousands in a day. Glass beads are also manufactured or moulded using a rotary machine where molten glass is fed in to the centre of a rotary mould and solid or hollow glass beads are formed.

The Bohemian glass industry was known for its ability to copy more expensive beads, and produced molded glass "lion's teeth", "coral", and "shells", which were popular in the 19th and early 20th century Africa trade.

Lampwork Beads

Lampworked dichroic glass bead showing thin film application.

A variant of the wound glass beadmaking technique, and a labor-intensive one, is what is traditionally called lampworking. In the Venetian industry, where very large quantities of beads were produced in the 19th century for the African trade, the core of a decorated bead was produced from molten glass at furnace temperatures, a large-scale industrial process dominated by men. The delicate multicolored decoration was then added by people, mostly women, working at home using an oil lamp or spirit lamp to re-heat the cores and the fine wisps of colored glass used to decorate them. These workers were paid on a piecework basis for the resulting lampwork beads. Modern lampwork beads are made by using a gas torch to heat a rod of glass and spinning the resulting thread around a metal rod covered in bead release. When the base bead has been formed, other colors of glass can be added to the surface to create many designs. After this initial stage of the beadmaking process, the bead can be further fired in a kiln to make it more durable.

Furnace glass beads.

Modern beadmakers use single or dual fuel torches, so `flameworked' is replacing the older term. Unlike a metalworking torch, or burner as some people in the trade prefer to call them, a flameworking torch is usually "surface mix"; that is, the oxygen and fuel (typically propane, though natural gas is also common) is mixed after it comes out of the torch, resulting in a quieter tool and less dirty flame. Also unlike metalworking, the torch is fixed, and the bead and glass move in the flame. American torches are usually mounted at about a 45 degree angle, a result of scientific glassblowing heritage; Japanese torches are recessed, and have flames coming straight up, like a large bunsen burner; Czech production torches tend to be positioned nearly horizontally.

Dichroic Glass Beads

Increasingly, dichroic glass is being used to produce high-end art beads. Dichroic glass has a thin film of metal fused to the surface of the glass, resulting in a surface that has a metallic sheen that changes between two colors when viewed at different angles. Beads can be pressed, or made with traditional lampworking techniques. If the glass is kept in the flame too long, the metallic coating will turn silver and burn off.

Furnace Glass

Italian glass blowing techniques such as latticinio and zanfirico are adapted here to make beads. Furnace glass uses large decorated canes built up out of smaller canes, encased in clear glass and then extruded to form the beads with linear and twisting stripe patterns. No air is blown into the glass. These beads require a large scale glass furnace and annealing kiln for manufacture.

Lead Crystal

Lead crystal beads are machine cut and polished. Their high lead content makes them sparkle more than other glass, but also makes them inherently fragile.

Other Methods

Lead glass (for neon signs) and, especially borosilicate is available in tubing, making true blown beads possible. (Soda-lime glass can be blown at the end of a metal tube, or, more commonly wound on the mandrel to make a hollow bead, but the former is unusual and the latter not a true mouth-blown technique.) In addition, beads can be fused from sheet glass or using ground glass.

Modern Ghana has an industry in beads molded from powdered glass. Also in Africa, Kiffa beads are made in Mauritania, historically by women, using powdered glass that the beadmaker usually grinds herself from commercially available glass seed beads and recycled glass.

Molded ground glass, if painted into the mold, is called pate de verre, and the technique can be used to make beads, though pendants and cabochons are more typical. Lampwork (and other) beads can be painted with glass paints.

Glass Fusing

Glass fusing is the joining together of pieces of glass at high temperature, usually in a kiln. This is usually done roughly between 700 °C (1,292 °F) and 820 °C (1,510 °F), and can range from *tack fusing* at lower temperatures, in which separate pieces of glass stick together but still retain their individual shapes, to *full fusing* at higher ones, in which separate pieces merge smoothly into one another.

Compatibility

Disparate pieces of glass must be compatible in order to ensure they can be fused properly. It is a common misconception that glasses having the same coefficient of expansion (COE) will be compatible. Coefficient of expansion is one indicator that glasses may be compatible, but there are many other factors that determine whether glasses are compatible. If incompatible glasses are fused together, it is unlikely that the fused piece will be able maintain structural integrity. The piece may shatter during the cooling process, or develop stress originating from the point of contact between the incompatible glasses over time, leading to fractures within the glass, and eventually breakage.

Generally, kiln-glass manufacturers will rate their glasses for compatibility with other glasses they make. In order to be certain that the glasses they use will be compatible, many glass fusers will adopt one manufacturer's glasses to use exclusively.

The stress in two pieces of incompatible glass that were fused can be observed by placing the item between two polarizing filters. This will show areas of tension which will develop stress and fracture over time.

Techniques

Most contemporary fusing methods involve stacking, or layering thin sheets of glass, often using different colors to create patterns or simple images. The stack is then placed inside the kiln (which is almost always electric, but can be heated by gas or wood) and then heated through a series of ramps (rapid heating) and soaks (holding the temperature at a specific point) until the separate pieces begin to bond together. The longer the kiln is held at the maximum temperature, the more thoroughly the stack will fuse, eventually softening and rounding the edges of the original shape. Once the desired effect has been achieved at the maximum desired temperature, the kiln temperature will be brought down quickly through the temperature range of 815 °C (1,499 °F) to 573 °C (1,063 °F) to avoid devitrification. The glass is then allowed to cool slowly over a specified time, soaking at specified temperature ranges which are essential to the annealing process. This prevents uneven cooling and breakage and produces a strong finished product.

This cooling takes place normally for a period of 10–12 hours in 3 stages.

The first stage- the rapid cool period is meant to place the glass into the upper end of the annealing range 516 °C (961 °F). The second stage- the anneal soak at 516 °C (961 °F) is meant to equalize the temperature at the core and the surface of the glass at 516 °C (961 °F) relieving the stress between those areas. The last stage, once all areas have had time to reach a consistent temperature, is the final journey to room temperature. The kiln is slowly brought down over the course of 2 hours to 371 °C (700 °F), soaked for 2 hours at 371 °C (700 °F), down again to 260 °C (500 °F) which ends the firing schedule. The glass will remain in the closed kiln until the pyrometer reads room temperature.

Note that these temperatures are not hard and fast rules. Depending on the kiln, the size of the project, the number of layers, the desired finished look, and even the brand of glass, ramp and soak temperatures and times may vary. Small pendants can be fired and cooled very rapidly. For instance, small glass pieces can be fired in as little as one hour.

Finished Products

Fused glass techniques are generally used to create art glass, glass tiles, and jewellery, notably beads. Slumping techniques allow the creation of larger, functional pieces like dishes, bowls, plates, and ashtrays. Producing functional pieces generally requires 2 or more separate firings, one to fuse the glass and a second slump it to shape.

Since the 1970s, more hobbyists have focused on using kiln-fused glass to make beads and components for jewellery. This has become especially popular since the introduction of glass manufactured for the specific purpose of fusing in a kiln.

Slumping

Slumping is a technique in which items are made in a kiln by means of shaping glass over molds at high temperatures. The slumping of a pyrometric cone is often used to measure temperature in a kiln.

Technique

Slumping glass is a highly technical operation that is subject to many variations, both controlled and uncontrolled. When an item is being slumped in a kiln, the mold over which it is being formed (which can be made of either ceramic, sand or metal) must be coated with a release agent that will stop the molten glass from sticking to the mold. Such release agents, a typical one being boron nitride, give off toxic fumes when they are first heated and must be used in a ventilated area.

The glass is cut to the shape of the mold (but slightly larger to allow for shrinkage) and placed on top of it, before the kiln is heated.

The stages of the firing can be varied but typically start to climb at quite a rapid rate until the heat places the glass in an "orange state" i.e., flexible. At that point, gravity will allow the glass to slump into the mold and the temperature is held at a constant for a period that is known as the "soak". Following this stage, the kiln is allowed to cool slowly so that the slumped glass can anneal and be removed from the kiln. If two differing colours of glass are used in a single piece of work, the same CoE (Coefficient of thermal Expansion) glass must be used, or the finished piece will suffer from fractures as the glass will shrink at differing rates and allow tension to build up to the point of destruction. To compensate for this, many glass manufacturers subscribe to make glass to the same CoE. Examples include Spectrum glass system 96 or uroborus 96 series, and the use of this glass will allow the cooling to remain uniform and ensure that no tension builds up as the work cools.

Fourcault Process

The Fourcault process is a method of manufacturing flat glass. First developed in Belgium by Émile Fourcault during the early 1900s, the process was used globally. Fourcault is an example of a "vertical draw" process, in that the glass is drawn against gravity in an upward direction. Gravity forces influence parts of the process.

Process

The Fourcault process requires a "pit" or drawing area and an assembly of machines to draw up the ribbon of glass while performing actions upon it that ensure desired quality and process yields. Today most glass manufacture has a "hot end" where the products are made. Fourcault is no exception.

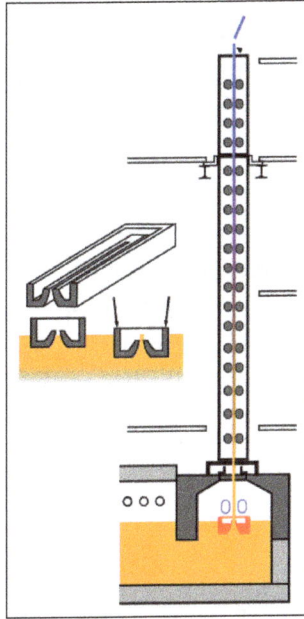

Fourcault drawing line with detail of the Debiteuse (red).

The action in Fourcault happens "at the draw", or area where the glass is taken from a liquid state into the start of the process needed to make it into flat glass.

At the bottom of the draw is the "pit" or place where the molten glass is sufficiently cooled to be close to forming temperature. The cooling process uses a device known as a "canal". As the name describes, a canal is a box shaped structure which conveys the glass from the refining area to the pit.

The canal links the pit with the "refining" area, a section of the glass furnace that removes gas bubbles and other sources of imperfection. Since refining requires much higher temperatures to release gas bubbles than those required to form the glass it is not possible to draw directly from the refining area, hence the need for canals.

Forming

The Fourcault Process uses a ceramic die to shape fused (or molten) glass into a ribbon of rectangular cross section. The die, known as a Debiteuse, floats in the molten glass inside of the pit to a prescribed depth which slightly pushes a part of the molten glass slightly above the top surface of the die. A slot is cut through the center of the Debiteuse, which is shaped to produce the best quality of glass.

The Debiteuse is the starting point of the vertical draw, where the glass begins to change from a hot syrupy mass into useful flat glass. We will call the glass from the point of the Debiteuse until it is cut a "ribbon".

The base of the ribbon is shielded from heat radiation from the fused glass so that it continues to hold the shape imparted to it by the Debiteuse. This cooling preserves the rectangular cross section of the drawn glass by cooling the ribbon glass below the temperature where it would collapse into a column or break back into the melted glass. It is especially important to shield the outside edges of the ribbon from heat so that they are firmer and will hold the rest of the ribbon in a proper shape. In some cases manufacturers will allow the edges to form thicker "bulbs", which are removed after final cutting.

Immediately after being drawn the ribbon is cooled using mechanical coolers so that it maintains its rectangular shape in two dimensions, but assumes a ribbon like structure that extends down into the Debi and upwards into a drawing assembly. This mechanical cooling allows the ribbon to hold its integrity. In the author's experience the mechanical coolers used water, contained in specially shaped radiators, to remove heat radiated by the ribbon.

Sometimes a mild vacuum is applied to the ribbon in this early part of the process since mechanical cooling can induce air currents which impact upon surface quality.

Quenching

Some manufacturers also will apply sulfur dioxide gas during the draw in order to change the chemistry of the glass on the surface. By changing the chemistry it is possible to affect the surface characteristics of the glass, improving its quality and durability.

Glass rollers hold the ribbon throughout various parts of the process, supporting its weight and continuing the drawing process.

The process continues as the ribbon is drawn upwards into a chimney like structure, where it is quenched or rapidly cooled. When the ribbon reaches the end of the process it is scored, or cut, and then removed for further processing in discrete sheets of flat glass.

The "bulb edges" are recycled as cullet (flawed glass which is remelted) or were resold for shelving or displays. Sometimes flawed parts of the sheets were removed, leaving behind decent quality flat glass.

Operations

Time, speed and spacing of the various phases of the process are critical factors in the Fourcault process. Fourcault process machine operators require experience in order to judge placement of the die, location of various parts of the process, and rates of draw. These must be balanced against glass quality and the age of the draw.

As the draw continues the glass in the pit grows cooler and cooler, eventually leading to failures or diminished quality. The draw must be stopped, the pit must be "heated back" and then the process can continue anew.

Glass chemistry has a huge impact on the process since it controls the melting, forming and annealing temperatures, liquidus temperature (point where various chemicals that make up the glass start to crystallize out of the glass) and rates of change of characteristics of the glass itself.

Occasionally the ribbon will break or crack, leading to failure of the drawing process. Such breaks, known as "checks" can be alleviated by using proper operating parameters. Sometimes an expedient measure, using a portable source of heat, can be used to make the checks migrate to the edge of the ribbon where they disappear. The author has even seen crude torches made of wood which can migrate the checks.

The resultant product is a form of flat glass which is suitable for lesser quality uses. Due to process instabilities Fourcault process glass can have waves, seeds (small gas bubbles) or stones (undissolved materials). This distorts the image seen through the glass. Fourcault glass is still being made as an architectural glass for historical restoration of buildings.

In terms of economics and product quality the Fourcault process has been supplanted in many countries by the Pilkington developed "Float" process. The Float process lets the molten glass settle on top of a pool of liquid tin, so that gravity creates a flat sheet. Due to various chemistry and physical aspects of window glass the Pilkington Float process produces a vastly superior product.

Lampworking

Lampworking is a type of glasswork whereby the artist utilizes a hot flame and various tools to wind and blow molten glass, usually onto special rods called mandrels, forming beads, figurines or other similar miniature artwork. Lampworking got its name from artists some time ago that used hot lamps to melt their glass. Modern day lampworking or flame working utilizes generally three types of fuel, either Mapp Gas, propane or a propane/oxygen mix. Forced air is also used in some applications to provide considerable heat.

A variety of torches are available to handle a wide range of lampworking techniques. For soft glass, (soda-lime glass) with a coefficient of expansion (COE) between 90 and 104, a special Propane or Mapp Gass torch can be used. Often this is beginners' first choice, since it is rather inexpensive. More specialized work requires a torch that mixes the propane with oxygen for a hotter flame. The most common type of torch that mixes both propane and oxygen and is used for soft glass is a "surface mix" torch. This type of torch does not mix the propane and oxygen until the gases exit the torch head. A torch that mixes the propane and oxygen internally (prior to the surface) is called a premix torch. Although some artists use a premix torch for soft glass, most use a surface mix torch because it is cooler than the premix torch.

However, a premix torch is more common with artists that work with a different type of glass called borosilicate glass (hard glass or boro) that has a COE of 33. There are some surface mix torches that do become hot enough to work with borosilicate glass, but a premix torch works best and is more common with boro users. There are also combination torches that have both premix and surface mix torch heads attached.

When lampworking, the artist must anneal their glass artwork, best done immediately after the artwork is complete. Annealing glass objects greatly reduces the chance that the glass will crack and break later. Annealing requires a kiln that can reach excessive temperatures. Check with your glass manufacturer or distributor to find which temperature is best for the glass you use. Also check the proper method for annealing. Each type of glass requires a different annealing schedule.

References

- Vitrification: newworldencyclopedia.org, Retrieved 18 June, 2019

- Cummings, K. 2002. A History of Glassforming. University of Pennsylvania Press ISBN 0812236475

- Ananthasayanam et.al. Final Shape of Precision Molded Optics: Part II—Validation and Sensitivity to Material Properties and Process Parameters, https://www.tandfonline.com/doi/abs/10.1080/01495739.2012.674838

- Lampworking: devardi.com, Retrieved 29 May, 2019

Applications

Glass has a wide range of applications in different fields of medicine and industry. A few of its medical applications are usage of glass in beaker, smartglasses, bioglass, bioactive glass, etc. and its industrial applications include the use of glass in mirror, windshield, liquid crystal display, etc. The topics elaborated in this chapter will help in gaining a better perspective about these applications of glass.

MEDICAL APPLICATIONS OF GLASS

Contact Lenses

Contact lenses are medical devices worn directly on the cornea of the eye. Like eyeglasses, contact lenses help to correct refractive errors and perform this function by adding or subtracting focusing power to the eye's cornea and lens. Contacts provide a safe and effective way to correct vision when used with care and proper supervision. They can offer a good alternative to eyeglasses, depending on your eyes and your lifestyle. Over 24 million people in the United States now wear contact lenses. For certain conditions, contact lenses may be considered medically necessary.

Cosmetic contact lenses are used to correct the same conditions that eyeglasses correct:

- Myopia (nearsightedness).
- Hyperopia (farsightedness).
- Astigmatism (distorted vision).
- Presbyopia (need for bifocals).

Types of Contact Lenses

Many types of contact lenses are available. The type of contacts you use depends on your particular situation. Your optometrist will be able to help you choose from the following types of lenses:

- Soft contact lenses: These are the most common type of contact lenses currently prescribed. These lenses are made materials that incorporate water, which makes them soft and flexible and allows oxygen to reach the cornea.

- ◦ Daily disposable lenses: Although generally more expensive, they carry a lower infection risk.

- ◦ Two week or monthly disposable lenses: For daily wear.

- ◦ Toric contact lenses: Correct moderate astigmatism.

- ◦ Bifocal contact lenses: Can be helpful for patients that need reading and distance correction but may not be right for everyone.

- • Gas-permeable lenses: These lenses are also known as "RGPs." They are rigid or "hard" lenses made of plastics combined with other materials—such as silicone and fluoropolymers—that allow oxygen in the air to pass directly through the lens. For this reason, they are called "gas-permeable."

Risk Factors

- • Daily-wear lenses should never be worn as extended-wear lenses. Misuse can lead to temporary and potentially sight threatening damage to the cornea. People who wear any type of lens overnight have a greater chance of developing infections of the cornea. These infections are often due to poor cleaning and lens care. Improper over wearing of contact lenses can result in intolerance, leading to the inability to wear contact lenses.

- • Gas-permeable lenses can potentially scratch the cornea if the lens does not fit properly or if the lens is worn while sleeping. They are also more likely to slide off the cornea and become hidden under the lid.

  - ◦ Gas-permeable lenses traditionally had a reputation for "popping out" of the eye. Newer lens designs have minimized the chance of losing a contact even during vigorous exercise.

  - ◦ Gas-permeable lenses and soft extended-wear contacts are the most likely to have protein build-up and cause lens-related allergies. Protein build-up results in discomfort, blurring, and intolerance to the lenses. Thus, nightly disinfection becomes imperative and you may need special cleaning solutions to dissolve the protein.

- • Rigid gas-permeable or disposable lenses may be good choices for someone with allergies.

Advantages and Disadvantages of Various Types of Contact Lenses

Rigid Gas-permeable

Made of slightly flexible plastics that allow oxygen to pass through to the eyes.

Advantages

- Excellent vision.
- Short adaptation period.
- Comfortable to wear.
- Correct most vision problems.
- Easy to put on and care for.
- Relatively long life.
- Available in tints (for handling purposes) and bifocals.
- Available for myopia control and corneal refractive therapy.

Disadvantages

- Require consistent wear to maintain adaptation.
- Can slip off center of eye more easily than other types.
- Debris can sometimes get under the lenses.
- Require regular office visits for follow-up care.

Daily-wear Soft Lenses

Made of soft, flexible plastic that allows oxygen to pass through to the eyes.

Advantages

- Very short adaptation period.
- More comfortable and more difficult to dislodge than RGP lenses.
- Available in tints and bifocals.
- Available in lenses that do not need to be cleaned.
- Great for active lifestyles.

Disadvantages

- Do not correct all vision problems.
- Vision may not be as sharp as with RGP lenses.
- Require regular office visits for follow-up care.
- Lenses wear out and must be replaced in a timely fashion.

Extended-wear

Available for Overnight Wear in Soft or RGP Lenses.

Advantages

- Can usually be worn up to seven days without removal.
- Some lenses are FDA-approved for up to 30 days.

Disadvantages

- Do not correct all vision problems.
- Require regular office visits for follow-up care.
- Could increase risk of complication.
- Require regular monitoring and professional care.

Extended-wear Disposable

Soft lenses worn for extended period of time (usually one to six days), then discarded. Lenses also available to wear from one to 30 days.

Advantages

- Require little or no cleaning.
- Lessened risk of eye infection if wearing instructions are followed.
- Available in tints and bifocals.
- Spare lenses available.

Disadvantages

- Vision may not be as sharp as with RGP lenses.
- Do not correct all vision problems.
- Handling may be more difficult.

Planned Replacement

Soft daily-wear lenses that are replaced on a planned schedule, most often either every two weeks, monthly or quarterly.

Advantages

- Simplified cleaning and disinfection.

- Good for eye health.

- Available in most prescriptions.

Disadvantages

- Vision may not be as sharp as with RGP lenses.

- Do not correct all vision problems.

- Handling may be more difficult.

Microscope Slide

A microscope slide is a thin flat piece of glass, typically 75 by 26 mm (3 by 1 inches) and about 1 mm thick, used to hold objects for examination under a microscope. Typically the object is mounted (secured) on the slide, and then both are inserted together in the microscope for viewing. This arrangement allows several slide-mounted objects to be quickly inserted and removed from the microscope, labeled, transported, and stored in appropriate slide cases or folders etc.

A set of standard 75 by 25 mm microscope slides.
The white area can be written on to label the slide.

Microscope slides are often used together with a cover slip or cover glass, a smaller and thinner sheet of glass that is placed over the specimen. Slides are held in place on the microscope's stage by slide clips, slide clamps or a cross-table which is used to achieve precise, remote movement of the slide upon the microscope's stage (such as in an automated / computer operated system, or where touching the slide with fingers is inappropriate either due to the risk of contamination or lack of precision).

Dimensions and Types

A standard microscope slide measures about 75 mm by 25 mm (3″ by 1″) and is about

1 mm thick. A range of other sizes are available for various special purposes, such as 75 x 50 mm for geological use, 46 x 27 mm for petrographic studies, and 48 x 28 mm for thin sections. Slides are usually made of common glass and their edges are often finely ground or polished.

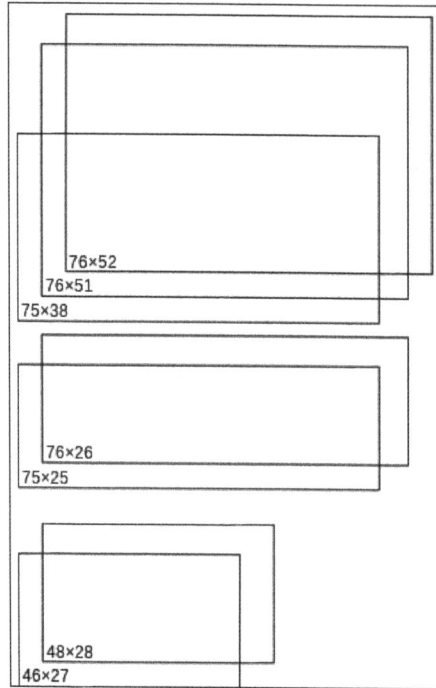

76×52
76×51
75×38

76×26
75×25

48×28
46×27

Common dimensions of microscope slides (in mm).

Microscope slides are usually made of optical quality glass, such as soda lime glass or borosilicate glass, but specialty plastics are also used. Fused quartz slides are often used when ultraviolet transparency is important, e.g. in fluorescence microscopy.

While plain slides are the most common, there are several specialized types. A concavity slide or cavity slide has one or more shallow depressions ("wells"), designed to hold slightly thicker objects, and certain samples such as liquids and tissue cultures. Slides may have rounded corners for increased safety or robustness, or a cut-off corner for use with a slide clamp or cross-table, where the slide is secured by a spring-loaded curved arm contacting one corner, forcing the opposing corner of the slide against a right angled arm which does not move. If this system were used with a slide which did not incorporate these cut-off corners, the corners would chip and the slide could shatter.

A graticule slide is marked with a grid of lines (for example, a 1 mm grid) that allows the size of objects seen under magnification to be easily estimated and provides reference areas for counting minute objects. Sometimes one square of the grid will itself be subdivided into a finer grid. Slides for specialized applications, such as hemocytometers for cell counting, may have various reservoirs, channels and barriers etched or ground on their upper surface. Various permanent markings or masks may be printed, sand-blasted, or deposited on the surface by the manufacturer, usually with inert materials such as PTFE.

Some slides have a frosted or enamel-coated area at one end, for labeling with a pencil or pen. Slides may have special coatings applied by the manufacturer, e.g. for chemical inertness or enhanced cell adhesion. The coating may have a permanent electric charge to hold thin or powdery samples. Common coatings include poly-L-lysine, silanes, epoxy resins, or even gold.

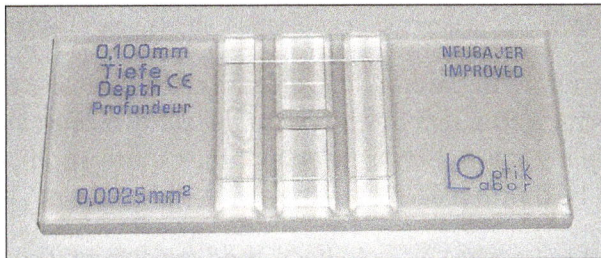

A Neubauer slide for cell counting.

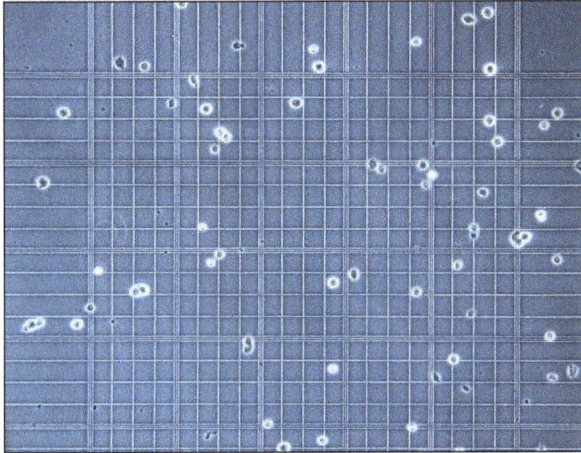

Microscope image of a Neubauer slide's graticule being used to count cells.

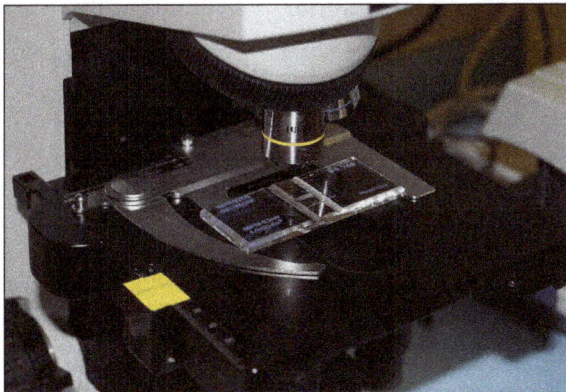

A Neubauer slide held in place on a microscope stand by a slide clamp on a cross-table.

Mounting

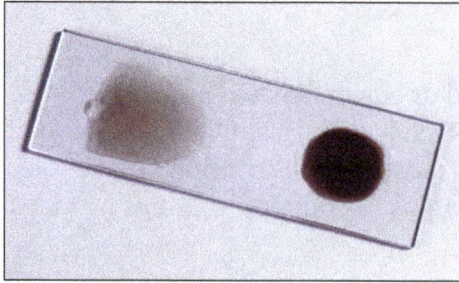

Blood smears for pathological examination, an example of wet mount.

The mounting of specimens on microscope slides is often critical for successful viewing. The problem has been given much attention in the last two centuries and is a well-developed area with many specialized and sometimes quite sophisticated techniques. Specimens are often held into place using the smaller glass cover slips.

The main function of the cover slip is to keep solid specimens pressed flat, and liquid samples shaped into a flat layer of even thickness. This is necessary because high-resolution microscopes have a very narrow region within which they focus.

The cover glass often has several other functions. It holds the specimen in place (either by the weight of the cover slip or, in the case of a wet mount, by surface tension) and protects the specimen from dust and accidental contact. It protects the microscope's objective lens from contacting the specimen and vice versa; in oil immersion microscopy or water immersion microscopy the cover slip prevents contact between the immersion liquid and the specimen. The cover slip can be glued to the slide so as to seal off the specimen, retarding dehydration and oxidation of the specimen and also preventing contamination. A number of sealants are in use, including commercial sealants, laboratory preparations, or even regular clear nail polish, depending on the sample. A solvent-free sealant that can be used for live cell samples is "valap", a mixture of vaseline, lanolin and paraffin in equal parts. Microbial and cell cultures can be grown directly on the cover slip before it is placed on the slide, and specimens may be permanently mounted on the slip instead of on the slide.

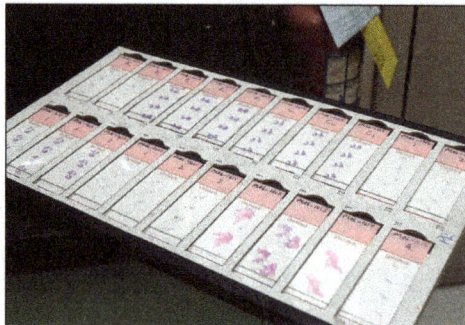

Microscope slides with prepared, stained, and labeled
tissue specimens in a standard 20-slide folder.

Cover slips are available in a range of sizes and thicknesses. Using the wrong thickness can result in spherical aberration and a reduction in resolution and image intensity. Specialty objectives may used to image specimens without coverslips, or may have correction collars that permit a user to accommodate for alternative coverslip thickness.

Dry Mount

In a dry mount, the simplest kind of mounting, the object is merely placed on the slide. A cover slip may be placed on top to protect the specimen and the microscope's objective and to keep the specimen still and pressed flat. This mounting can be successfully used for viewing specimens like pollen, feathers, hairs, etc. It is also used to examine particles caught in transparent membrane filters (e.g., in analysis of airborne dust).

Wet Mount or Temporary Mount

In a wet mount, the specimen is placed in a drop of water or other liquid held between the slide and the cover slip by surface tension. This method is commonly used, for example, to view microscopic organisms that grow in pond water or other liquid media, especially when studying their movement and behavior. Care must be taken to exclude air bubbles that would interfere with the viewing and hamper the organisms' movements. An example of a temporary wet mount is a lactofuchsin mount, which provides both a sample mounting, as well as a fuchsine staining.

Prepared Mount or Permanent Mount

For pathological and biological research, the specimen usually undergoes a complex histological preparation that involves fixing it to prevent decay, removing any water contained in it, replacing the water with paraffin, cutting it into very thin sections using a microtome, placing the sections on a microscope slide, staining the tissue using various stains to reveal specific tissue components, clearing the tissue to render it transparent and covering it with a coverslip and mounting medium.

Strewn Mount

Strewn mounting describes the production of palynological microscope slides by suspending a concentrated sample in distilled water, placing the samples on a slide, and allowing the water to evaporate.

Mounting Media

The mounting medium is the solution in which the specimen is embedded, generally under a cover glass. Simple liquids like water or glycerol can be considered mounting media, though the term generally refers to compounds that harden into a permanent mount. Popular mounting media include *Permount*, and Hoyer's mounting medium and an alternative *glycerine jelly* Properties of a good mounting medium include having a refractive index close to that of glass (1.518), non-reactivity with the specimen, stability over time without crystallizing, darkening, or changing refractive index, solubility in the medium the specimen was prepared in (either aqueous or non-polar, such as xylene or toluene), and not causing the specimen stain to fade or leach.

Examples of Mounting Media

Aqueous

Popularly used in immunofluorescent cytochemistry where the fluorescence cannot be archived. The temporary storage must be done in a dark moist chamber. Common examples are:

- Glycerol-PBS (9:1) with antiquench, e.g. any of the following:
  - p-phenylenediamine.
  - propyl gallate.
  - 1,4-Diazabicyclo (2,2,2)-octane (DABCO) (very popular).
  - Ascorbic acid.
  - Mowiol or Gelvatol.
- Gelatin.
- Mount.
- Vectashield.
- Prolong Gold.
- CyGEL / CyGEL Sustain (to immobilize living, unfixed cells and organisms).

Non-aqueous

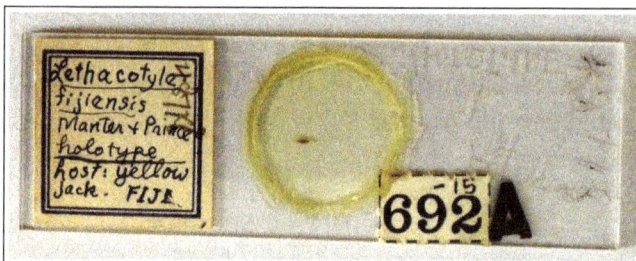

Slide of 60-year-old holotype specimen of a flatworm
(Lethacotyle fijiensis) permanently mounted in Canada balsam.

Used when a permanent mount is required:

- Permount (toluene and a polymer of a-pinene, b-pinene, dipentene, b-phellandrene).

- Canada balsam.

- DPX (*Distrene 80* – a commercial polystyrene, a *p*lasticizer e.g. dibutyl phthalate and xylene).

- DPX new (with xylene but free of carcinogenic dibutyl phthalate).

- Entellan (with toluene).

- Entellan new.

- Hempstead Halide Hoyer's Medium (a proprietary formulation of the traditional Hoyer's medium containing 60% Chloral, but free of known carcinogens).

- Neo-Mount (compatible with aliphatic neo-clear but not compatible with aromatic solvents like xylene).

Borosilicate Glass

Borosilicate glass is a type of glass with silica and boron trioxide as the main glass-forming constituents. Borosilicate glasses are known for having very low coefficients of thermal expansion ($\approx 3 \times 10^{-6}$ K^{-1} at 20 °C), making them more resistant to thermal shock than any other common glass. Such glass is subjected to less thermal stress and can withstand temperature differentials without fracturing of about 165 °C (329 °F). It is commonly used for the construction of reagent bottles. Borosilicate glass is sold under various trade names, including Borcam, Borosil, DURAN, Pyrex, Supertek, Suprax, Simax, BSA 60, BSC 51 (by NIPRO), Heatex, Endural, Schott, Refmex, Kimble, and MG.

Glass made of borosilicated glass.

Manufacturing Process

Borosilicate glass is created by combining and melting boric oxide, silica sand, soda ash, and alumina. Since borosilicate glass melts at a higher temperature than ordinary silicate glass, some new techniques were required for industrial production. The manufacturing process depends on the product geometry and can be differentiated between different methods like floating, tube drawing, or moulding.

Physical Characteristics

The common type of borosilicate glass used for laboratory glassware has a very low thermal expansion coefficient (3.3×10^{-6} K^{-1}), about one-third that of ordinary soda-lime glass. This reduces material stresses caused by temperature gradients, which makes borosilicate a more suitable type of glass for certain applications. Fused quartzware is even better in this respect (having one-fifteenth the thermal expansion of soda-lime glass); however, the difficulty of working with fused quartz makes quartzware much more expensive, and borosilicate glass is a low-cost compromise. While more resistant to thermal shock than other types of glass, borosilicate glass can still crack or shatter when subjected to rapid or uneven temperature variations.

Among the characteristic properties of this glass family are:

- Different borosilicate glasses cover a wide range of different thermal expansions, enabling direct seals with various metals and alloys like molybdenum glass with a CTE of 4,6, tungsten with a CTE around 4,0 and Kovar with a CTE around 5,0 because of the matched CTE with the sealing partner.

- Allowing high maximum temperatures of typically about 500 °C (932 °F).

- Showing an extremely high chemical resistance in corrosive environments. Norm tests for example for acid resistance create extreme conditions and reveal very low impacts on glass.

The softening point (temperature at which viscosity is approximately $10^{7.6}$ poise) of type 7740 Pyrex is 820 °C (1,510 °F).

Borosilicate glass is less dense (about 2.23 g/cm³) than typical soda-lime glass due to the low atomic mass of boron. Its mean specific heat capacity at constant pressure (20–100 °C) is 0.83 J/(g·K), roughly one fifth of water's.

The temperature differential that borosilicate glass can withstand before fracturing is about 165 °C (329 °F). This compares well with soda lime glass, which can withstand only a 37 °C (99 °F) change in temperature and is why typical kitchenware made from traditional soda-lime glass will shatter if a vessel containing boiling water is placed on ice, but Pyrex or other borosilicate laboratory glass will not.

Optically, borosilicate glasses are crown glasses with low dispersion (Abbe numbers around 65) and relatively low refractive indices (1.51–1.54 across the visible range).

Glass Families

For the purposes of classification, borosilicate glass can be roughly arranged in the following groups, according to their oxide composition (in mass fractions). Characteristic of borosilicate glasses is the presence of substantial amounts of silica (SiO_2) and boric oxide (B_2O_3, >8%) as glass network formers. The amount of boric oxide affects the glass properties in a particular way. Apart from the highly resistant varieties (B_2O_3 up to a maximum of 13%), there are others that – due to the different way in which the boric oxide is incorporated into the structural network – have only low chemical resistance (B_2O_3 content over 15%). Hence we differentiate between the following subtypes.

Non-alkaline-earth Borosilicate Glass (Borosilicate Glass 3.3)

The B_2O_3 content for borosilicate glass is typically 12–13% and the SiO_2 content over 80%. High chemical durability and low thermal expansion (3.3×10^{-6} K^{-1}) – the lowest of all commercial glasses for large-scale technical applications – make this a multitalented glass material. High-grade borosilicate flat glasses are used in a wide variety of industries, mainly for technical applications that require either good thermal resistance, excellent chemical durability, or high light transmission in combination with a pristine surface quality. Other typical applications for different forms of borosilicate glass include glass tubing, glass piping, glass containers, etc. especially for the chemical industry.

Alkaline-earth-containing Borosilicate Glasses

In addition to about 75% SiO_2 and 8–12% B_2O_3, these glasses contain up to 5% alkaline earths and alumina (Al_2O_3). This is a subtype of slightly softer glasses, which have thermal expansions in the range $(4.0–5.0) \times 10^{-6}$ K^{-1}.

High-borate Borosilicate Glasses

Glasses containing 15–25% B_2O_3, 65–70% SiO_2, and smaller amounts of alkalis and Al_2O_3 as additional components have low softening points and low thermal expansion. Sealability to metals in the expansion range of tungsten and molybdenum and high electrical insulation are their most important features. The increased B_2O_3 content reduces the chemical resistance; in this respect, high-borate borosilicate glasses differ widely from non-alkaline-earth and alkaline-earth borosilicate glasses. Among these are also borosilicate glasses that transmit UV light down to 180 nm, which combine the best of the borosilicate glass and the quartz world.

Usage

Borosilicate glass has a wide variety of uses ranging from cookware to lab equipment, as well as a component of high-quality products such as implantable medical devices and devices used in space exploration.

Health and Science

Virtually all modern laboratory glassware is made of borosilicate glass. It is widely used in this application due to its chemical and thermal resistance and good optical clarity, but the glass can react with sodium hydride upon heating to produce sodium borohydride, a common laboratory reducing agent. Fused quartz is also found in some laboratory equipment when its higher melting point and transmission of UV are required (e.g. for tube furnace liners and UV cuvettes), but the cost and difficulty of working with quartz make it excessive for the majority of laboratory equipment.

Borosilicate beakers.

Additionally, borosilicate tubing is used as the feedstock for the production of parenteral drug packaging, such as vials and pre-filled syringes, as well as ampoules and dental cartridges. The chemical resistance of borosilicate glass minimizes the migration of sodium ions from the glass matrix, thus making it well suited for injectable-drug applications. This type of glass is typically referred to as USP / EP JP Type I.

Borosilicate is widely used in implantable medical devices such as prosthetic eyes, artificial hip joints, bone cements, dental composite materials (white fillings) and even in breast implants.

Many implantable devices benefit from the unique advantages of borosilicate glass encapsulation. Applications include veterinary tracking devices, neurostimulators for the treatment of epilepsy, implantable drug pumps, cochlear implants, and physiological sensors.

Electronics

During the mid-20th century, borosilicate glass tubing was used to pipe coolants (often distilled water) through high-power vacuum-tube–based electronic equipment, such as commercial broadcast transmitters. It was also used for the envelope material for glass transmitting tubes which operated at high temperatures.

Borosilicate glasses also have an application in the semiconductor industry in the development of microelectromechanical systems (MEMS), as part of stacks of etched silicon wafers bonded to the etched borosilicate glass.

Cookware

Arc International bakeware.

Glass cookware is another common usage. Borosilicate glass is used for measuring cups, featuring screen printed markings providing graduated measurements. Borosilicate glass is sometimes used for high-quality beverage glassware. Borosilicate glass is thin and durable, microwave- and dishwasher-safe.

Lighting

Many high-quality flashlights use borosilicate glass for the lens. This increases light transmittance through the lens compared to plastics and lower-quality glass.

Several types of high-intensity discharge (HID) lamps, such as mercury-vapor and metal-halide lamps, use borosilicate glass as the outer envelope material.

New lampworking techniques led to artistic applications such as contemporary glass marbles. The modern studio glass movement has responded to color. Borosilicate is commonly used in the glassblowing form of lampworking and the artists create a range of products such as jewelry, kitchenware, sculpture, as well as for artistic glass smoking pipes.

Lighting manufacturers use borosilicate glass in their refractors.

Organic light-emitting diode (for display and lighting purposes) also uses borosilicate glass (BK7). The thicknesses of the BK7 glass substrates are usually less than 1 millimeter for the OLED fabrication. Due to its optical and mechanical characteristics in relation with cost, BK7 is a common substrate in OLEDs. However, depending on the application, soda-lime glass substrates of similar thicknesses are also used in OLED fabrication.

Optics

Many astronomical reflecting telescopes use glass mirror components made of borosilicate glass because of its low coefficient of thermal expansion. This makes very precise optical surfaces possible that change very little with temperature, and matched glass mirror components that "track" across temperature changes and retain the optical system's characteristics.

The Hale Telescope's 200 inch mirror is made of borosilicate glass.

The optical glass most often used for making instrument lenses is Schott BK-7 (or the equivalent from other makers), a very finely made borosilicate crown glass. It is also designated as 517642 glass after its 1.517 refractive index and 64.2 Abbe number. Other less costly borosilicate glasses, such as Schott B270 or the equivalent, are used to make "crown-glass" eyeglass lenses. Ordinary lower-cost borosilicate glass, like that used to make kitchenware and even reflecting telescope mirrors, cannot be used for high-quality lenses because of the striations and inclusions common to lower grades of this type of glass. The maximal working temperature is 268 °C (514 °F). While it transitions to a liquid starting at 288 °C (550 °F) (just before it turns red-hot), it is not workable until it reaches over 538 °C (1,000 °F). That means that in order to industrially produce this glass, oxygen/fuel torches must be used. Glassblowers borrowed technology and techniques from welders.

Rapid Prototyping

Borosilicate glass has become the material of choice for fused deposition modeling (FDM), or fused filament fabrication (FFF), build plates. Its low coefficient of expansion makes borosilicate glass, when used in combination with resistance-heating plates and pads, an ideal material for the heated build platform onto which plastic materials are extruded one layer at a time. The initial layer of build must be placed onto a substantially flat, heated surface to minimize shrinkage of some build materials (ABS, polycarbonate, polyamide, etc.) due to cooling after deposition. The build plate will cycle from room temperature to between 100 °C and 130 °C for each prototype that is built. The temperature, along with various coatings (Kapton tape, painter tape, hair spray, glue stick, ABS+acetone slurry, etc.), ensure that the first layer may be adhered to and remain adhered to the plate, without warping, as the first and subsequent layers cool following extrusion. Subsequently, following the build, the heating elements and plate are allowed to cool. The resulting residual stress formed when the plastic contracts as it cools, while the glass remains relatively dimensionally unchanged due to the low coefficient of thermal expansion, provides a convenient aid in removing the otherwise mechanically bonded plastic from the build plate. In some cases the parts self-separate as the developed stresses overcome the adhesive bond of the build material to the coating material and underlying plate.

Other

Aquarium heaters are sometimes made of borosilicate glass. Due to its high heat resistance, it can tolerate the significant temperature difference between the water and the nichrome heating element.

Specialty glass smoking pipes for cannabis and tobacco are made from borosilicate glass. The high heat resistance makes the pipes more durable. Some harm reduction organizations also give out borosilicate pipes intended for smoking crack cocaine, as the heat resistance prevents the glass from cracking, causing cuts and burns that can spread hepatitis C.

Most premanufactured glass guitar slides are made of borosilicate glass.

Borosilicate is also a material of choice for evacuated-tube solar thermal technology because of its high strength and heat resistance.

The thermal insulation tiles on the Space Shuttle were coated with a borosilicate glass.

Borosilicate glasses are used for immobilisation and disposal of radioactive wastes. In most countries high-level radioactive waste has been incorporated into alkali borosilicate or phosphate vitreous waste forms for many years; vitrification is an established technology. Vitrification is a particularly attractive immobilization route because of the high chemical durability of the vitrified glass product. This characteristic has been exploited by the industry for centuries. The chemical resistance of glass can allow it to remain in a corrosive environment for many thousands or even millions of years.

Borosilicate glass tubing is used in specialty TIG welding torch nozzles in place of standard alumina nozzles. This allows a clear view of the arc in situations where visibility is limited.

Trade Names

Borosilicate glass is offered in slightly different compositions under different trade names:

- Borofloat of Schott AG, a borosilicate glass, which is produced to flat glass in a float process.

- BK7 of Schott, a borosilicate glass with a high level of purity. Main use in lens and mirrors for laser, cameras and telescopes.

- Duran of DURAN Group, similar to Pyrex, Simax or Jenaer Glas.

- Fiolax of Schott, mainly used for containers for pharmaceutical applications.

- Ilmabor of TGI [de] (2014 insolvency), mainly used for containers and equipment in laboratories and medicine.

- Jenaer Glas of Zwiesel Kristallglas, formerly Schott AG. Mainly used for kitchenware.

- Rasotherm of VEB Jenaer Glaswerk Schott & Genossen, for technical glass.

- Simax of Kavalierglass a.s., Czechia, produced for both laboratory and consumer markets.

- Willow Glass is an alkali free, thin and flexible borosilicate glass of Corning.

Borosilicate Nanoparticles

It was initially thought that borosilicate glass could not be formed into nanoparticles, since an unstable boron oxide precursor prevented successful forming of these shapes. However, in 2008 a team of researchers from the Swiss Federal Institute of Technology at Lausanne were successful in forming borosilicate nanoparticles of 100 to 500 nanometers in diameter. The researchers formed a gel of tetraethylorthosilicate and trimethoxyboroxine. When this gel is exposed to water under proper conditions, a dynamic reaction ensues which results in the nanoparticles.

In Lampworking

Borosilicate (or "boro", as it is often called) is used extensively in the glassblowing process lampworking; the glassworker uses a burner torch to melt and form glass, using a variety of metal and graphite tools to shape it. Borosilicate is referred to as "hard glass" and has a higher melting point (approximately 3,000 °F / 1648 °C) than "soft glass", which is preferred for glassblowing by beadmakers. Raw glass used in lampworking comes in glass rods for solid work and glass tubes for hollow work tubes and vessels/containers. Lampworking is used to make complex and custom scientific apparatus; most major universities have a lampworking shop to manufacture and repair their glassware. For this kind of "scientific glassblowing", the specifications must be exact and the glassblower must be highly skilled and able to work with precision. Lampworking is also done as art, and common items made include goblets, paper weights, pipes, pendants, compositions and figurines.

In 1968, English metallurgist John Burton brought his hobby of hand-mixing metallic oxides into borosilicate glass to Los Angeles. Burton began a glass workshop at Pepperdine College, with instructor Margaret Youd. A few of the students in the classes, including Suellen Fowler, discovered that a specific combination of oxides made a glass that would shift from amber to purples and blues, depending on the heat and flame atmosphere. Fowler shared this combination with Paul Trautman, who formulated the first small-batch colored borosilicate recipes. He then founded Northstar Glassworks in the mid-1980s, the first factory devoted solely to producing colored borosilicate glass rods and tubes for use by artists in the flame. Trautman also developed the techniques and technology to make the small-batch colored boro that is used by a number of similar companies.

Beadmaking

In recent years, with the resurgence of lampworking as a technique to make handmade glass beads, borosilicate has become a popular material in many glass artists' studios. Borosilicate for beadmaking comes in thin, pencil-like rods. Glass Alchemy, Trautman Art Glass, and Northstar are popular manufacturers, although there are other brands available. The metals used to color borosilicate glass, particularly silver, often create strikingly beautiful and unpredictable results when melted in an oxygen-gas torch flame. Because it is more shock-resistant and stronger than soft glass, borosilicate is particularly suited for pipe making, as well as sculpting figures and creating large beads. The tools used for making glass beads from borosilicate glass are the same as those used for making glass beads from soft glass.

Laboratory Flask

Laboratory flasks are vessels or containers that fall into the category of laboratory equipment known as glassware. In laboratory and other scientific settings, they are usually referred to simply as flasks. Flasks come in a number of shapes and a wide range of sizes, but a common distinguishing aspect in their shapes is a wider vessel "body" and one (or sometimes more) narrower tubular sections at the top called necks which have an opening at the top. Laboratory flask sizes are specified by the volume they can hold, typically in metric units such as milliliters (mL or ml) or liters (L or l). Laboratory flasks have traditionally been made of glass, but can also be made of plastic.

At the opening at top of the neck of some glass flasks such as round-bottom flasks, retorts, or sometimes volumetric flasks, there are outer (or female) tapered (conical) ground glass joints. Some flasks, especially volumetric flasks, come with a laboratory rubber stopper, bung, or cap for capping the opening at the top of the neck. Such stoppers can be made of glass or plastic. Glass stoppers typically have a matching tapered inner (or male) ground glass joint surface, but often only of stopper quality. Flasks which do not come with such stoppers or caps included may be capped with a rubber bung or cork stopper.

Flasks can be used for making solutions or for holding, containing, collecting, or sometimes volumetrically measuring chemicals, samples, solutions, etc. for chemical reactions or other processes such as mixing, heating, cooling, dissolving, precipitation, boiling (as in distillation), or analysis.

List of Flasks

There are several types of laboratory flasks, all of which have different functions within the laboratory. Flasks, because of their use, can be divided into:

- Reaction flasks, which are usually spherical (i.e. round-bottom flask) and are accompanied by their necks, at the ends of which are ground glass joints to quickly and tightly connect to the rest of the apparatus (such as a reflux condenser or dropping funnel). The reaction flask is usually made of thick glass and they can tolerate large pressure differences, with the result that you can keep two both in a reaction under vacuum, and pressure, sometimes simultaneously. Some varieties are:
  - Multiple neck flasks, which can have two to five, or less commonly, six necks, each topped by ground glass connections are used in more complex reactions that require the controlled mixing of multiple reagents.

○ Schlenk flask is a spherical flask with a ground glass opening and a hose outlet with a vacuum stopcock. The tap makes it easy to connect the flask to a vacuum-nitrogen line through the hose and carrying out the reaction either in vacuum or in atmosphere nitrogen.

- Distillation flasks, which are intended to contain mixtures, which are subject to distillation, as well as to receive the products of distillation, distillation flasks are available in various shapes. Similar to the reaction flask, the distillation flasks usually have only one narrow neck and a ground glass joint and are made of thinner glass than the reaction flask, so that it is easier to heat. They are sometimes spherical, test tube shaped, or pear-shaped, also known as a Kjeldahl Flask, due to its use with Kjeldahl bulbs.

- Reagent flasks are usually a flat-bottomed flask, which can thus be conveniently placed on the table or in a cabinet. These flasks cannot withstand too much pressure or temperature differences, due to the stresses which arise in a flat bottom, these flasks are usually made of weaker glass than reaction flasks. Certain types of flasks are supplied with a ground glass stopper in them, and others that are threaded neck and closes with an appropriate nut or automatic dispenser, these flasks are available in two standard shapes.

- Round-bottom flasks are shaped like a tube emerging from the top of a sphere. The flasks are often long neck; sometimes they have the incision on the neck, which precisely defines the volume of flask. They can be used in distillations, or in the heating a product. These types of flask are alternatively called Florence flasks.

- Flasks with flat bottom.

- Cassia flasks, for the analysis of essential oils and aldehyde determination, approx. 100 ml, neck graduated 0 - 6 : 0.1 ml.

- Erlenmeyer flask [introduced in 1861 by German chemist Emil Erlenmeyer] - is shaped like a cone, usually completed by the ground joint, the conical flasks are very popular because of their low price (they are easy to manufacture) and portability.

- Volumetric flask is used for preparing liquids with volumes of high precision. It is a flask with an approximately pear-shaped body and a long neck with a circumferential fill line.

- Dewar flask is a double-walled flask having a near-vacuum between the two walls. These come in a variety of shapes and sizes; some are large and tube-like, others are shaped like regular flasks.

- Evaporating flasks (for rotary evaporator) centered, pear shaped, with socket or with flange.

- Powder flasks, for drying of powdered substances, pear shaped, with socket.

- Retorts are simplified distillation apparatuses, with long, down turned necks, and round bases. They have largely been replaced by condensers.

- Büchner flask or Sidearm flask or Suction flask - they are a flat-bottomed flask, but made of very thick and resistant glass. They are usually a cone shape - similar to the shape of an Erlenmeyer flask, but also have side neck, usually affixed to the side, 2 / 3 up from the bottom. The flasks are used to cooperate with vacuum aspirator or vacuum pumps in the vacuum filtration, or as additional security during the distillation and other processes carried out under reduced pressure.

- Culture flasks for growing cells are designed to improve aeration by including baffles that aid in mixing when placed on a shaker table.

- Beaker (glassware).

Many of these flasks can be wrapped in a protective outer layer of glass, leaving a gap between the inner and outer walls. These are called *jacketed flasks*; they are often used in a reaction using a cooling fluid.

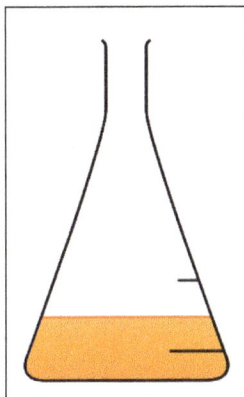

Erlenmeyer flask or conical flask.

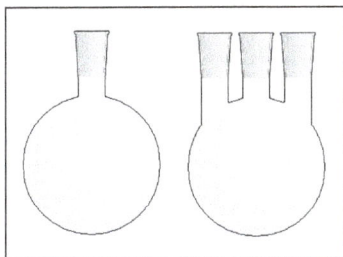

Round-bottom flask — a flask with a spherical body and one or more necks with ground glass joints.

Beaker

A beaker is generally a cylindrical container with a flat bottom. Most also have a small spout (or "beak") to aid pouring, as shown in the picture. Beakers are available in a wide range of sizes, from one millilitre up to several litres. A beaker is distinguished from a flask by having straight rather than sloping sides. The exception to this definition is a slightly conical-sided beaker called a Philips beaker.

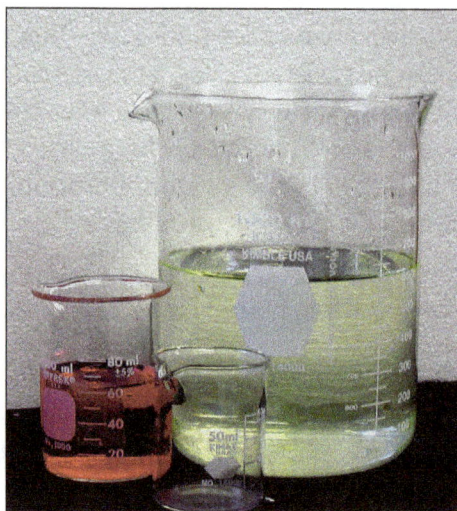

Beakers of several sizes.

Beakers are commonly made of glass (today usually borosilicate glass), but can also be in metal (such as stainless steel or aluminium) or certain plastics (notably polythene, polypropylene, PTFE). A common use for polypropylene beakers is gamma spectral analysis of liquid and solid samples.

Construction and Use

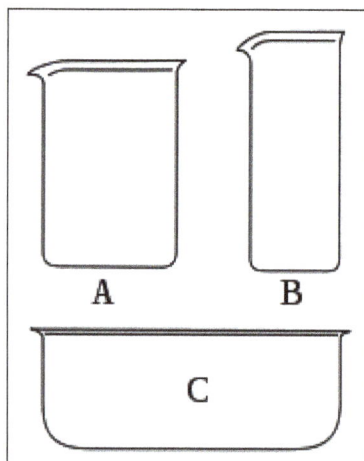

(A) A low-form or Griffin form beaker (B) A tall-form
or Berzelius beaker(C) A flat beaker or crystallizer.

Standard or "low-form" (A) beakers typically have a height about 1.4 times the diameter. The common low form with a spout was devised by John Joseph Griffin and is therefore sometimes called a Griffin beaker. These are the most universal character and are used for various purposes—from preparing solutions and decanting supernatant fluids to holding waste fluids prior to disposal to performing simple reactions. Low form beakers are likely to be used in some way when performing a chemical experiment. "Tall-form" (B) beakers have a height about twice their diameter. These are sometimes called Berzelius beakers and are mostly used for titration. Flat beakers (C) are often called "crystallizers" because most are used to perform crystallization, but they are also often used as a vessel for use in hot-bath heating. These beakers usually do not have a flat scale.

The presence of a spout means that the beaker cannot have a lid. However, when in use, beakers may be covered by a watch glass to prevent contamination or loss of the contents, but allowing venting via the spout. Alternatively, a beaker may be covered with another larger beaker that has been inverted, though a watch glass is preferable.

Beakers are often *graduated*, that is, marked on the side with lines indicating the volume contained. For instance, a 250 mL beaker might be marked with lines to indicate 50, 100, 150, 200, and 250 mL of volume. These marks are not intended for obtaining a precise measurement of volume (a graduated cylinder or a volumetric flask would be a more appropriate instrument for such a task), but rather an estimation. Most beakers are accurate to within ~10%.

Smartglasses

Smartglasses or smart glasses are wearable computer glasses that add information alongside or to what the wearer sees. Alternatively, smartglasses are sometimes defined as wearable computer glasses that are able to change their optical properties at runtime. Smart sunglasses which are programmed to change tint by electronic means are an example of the latter type of smartglasses.

Using the touch pad built on the side of the 2013 Google Glass to communicate with the user's phone using Bluetooth.

Superimposing information onto a field of view is achieved through an optical head-mounted display (OHMD) or embedded wireless glasses with transparent heads-up display (HUD) or augmented reality (AR) overlay. These systems have the capability to reflect projected digital images as well as allow the user to see through it or see better with it. While early models can perform basic tasks, such as serving as a front end display for a remote system, as in the case of smartglasses utilizing cellular technology or Wi-Fi, modern smart glasses are effectively wearable computers which can run self-contained mobile apps. Some are handsfree and can communicate with the Internet via natural language voice commands, while others use touch buttons.

Like other computers, smartglasses may collect information from internal or external sensors. It may control or retrieve data from other instruments or computers. It may support wireless technologies like Bluetooth, Wi-Fi, and GPS. A small number of models run a mobile operating system and function as portable media players to send audio and video files to the user via a Bluetooth or WiFi headset. Some smartglasses models also feature full lifelogging and activity tracker capability.

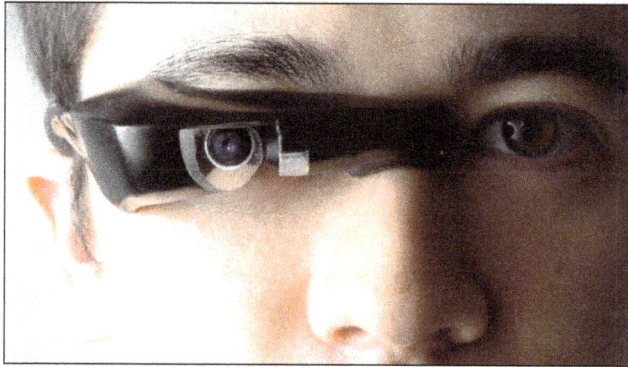

Man wearing a 1998 EyeTap, Digital Eye Glass.

Smartglasses devices may also have features found on a smartphone. Some have activity tracker functionality features (also known as *"fitness tracker"*) as seen in some GPS watches.

Features and Applications

As with other lifelogging and activity tracking devices, the GPS tracking unit and digital camera of some smartglasses can be used to record historical data. For example, after the completion of a workout, data can be uploaded into a computer or online to create a log of exercise activities for analysis. Some smart watches can serve as full GPS navigation devices, displaying maps and current coordinates. Users can "mark" their current location and then edit the entry's name and coordinates, which enables navigation to those new coordinates.

Although some smartglasses models manufactured in the 21st century are completely functional as standalone products, most manufacturers recommend or even require that consumers purchase mobile phone handsets that run the same operating system so that the two devices can be synchronized for additional and enhanced functionality. The smartglasses can work as an extension, for head-up display (HUD) or remote control of the phone and alert the user to communication data such as calls, SMS messages, emails, and calendar invites.

Security Applications

Smart glasses could be used as a body camera. In 2018, Chinese police in Zhengzhou and Beijing were using smart glasses to take photos which are compared against a government database using facial recognition to identify suspects, retrieve an address, and track people moving beyond their home areas.

Healthcare Applications

Several proofs of concept for Google Glasses have been proposed in healthcare. In July 2013, Lucien Engelen started research on the usability and impact of Google Glass in health care. Engelen, who is based at Singularity University and in Europe at Radboud University Medical Center, is participating in the Glass Explorer program.

Key findings of Engelen's research included:

- The quality of pictures and video are usable for healthcare education, reference, and remote consultation. The camera needs to be tilted to different angle for most of the operative procedures.

- Tele-consultation is possible—depending on the available bandwidth—during operative procedures.

- A stabilizer should be added to the video function to prevent choppy transmission when a surgeon looks to screens or colleagues.

- Battery life can be easily extended with the use of an external battery.

- Controlling the device and programs from another device is needed for some features because of a sterile environment.

- Text-to-speech ("Take a Note" to Evernote) exhibited a correction rate of 60 percent, without the addition of a medical thesaurus.

- A protocol or checklist displayed on the screen of Google Glass can be helpful during procedures.

Display Types

Various techniques have existed for see-through HMDs. Most of these techniques can be summarized into two main families: "Curved Mirror" (or Curved Combiner) based and "Waveguide" or "Light-guide" based. The mirror technique has been used in

EyeTaps, by Meta in their Meta 1, by Vuzix in their Star 1200 product, by Olympus, and by Laster Technologies.

Various waveguide techniques have existed for some time. These techniques include diffraction optics, holographic optics, polarized optics, reflective optics, and projection:

- Diffractive waveguide: Slanted diffraction grating elements (nanometric 10E-9). Nokia technique now licensed to Vuzix.

- Holographic waveguide: 3 holographic optical elements (HOE) sandwiched together (RGB). Used by Sony and Konica Minolta.

- Reflective waveguide: Thick light guide with single semi-reflective mirror. This technique is used by Epson in their Moverio product.

- Virtual retinal display (VRD): Also known as a retinal scan display (RSD) or retinal projector (RP), is a display technology that draws a raster display (like a television) directly onto the retina of the eye - developed by MicroVision, Inc.

The Technical Illusions castAR uses a different technique with clear glass. The glasses have a projector, and the image is returned to the eye by a reflective surface.

Smart Sunglasses

Smart sunglasses which are able to change their light filtering properties at runtime generally use liquid crystal technology. As lighting conditions change, for example when the user goes from indoors to outdoors, the brightness ratio also changes and can cause undesirable vision impairment. An attractive solution for overcoming this issue is to incorporate dimming filters into smart sunglasses which control the amount of ambient light reaching the eye. An innovative liquid crystal based component for use in the lenses of smart sunglasses is PolarView by LC-Tec. PolarView offers analog dimming control, with the level of dimming being adjusted by an applied drive voltage.

Another type of smart sunglasses uses adaptive polarization filtering (ADF). ADF-type smart sunglasses can change their polarization filtering characteristics at runtime. For example, ADF-type smart sunglasses can change from horizontal polarization filtering to vertical polarization filtering at the touch of a button.

The lenses of smart sunglasses can be manufactured out of multiple adaptive cells, therefore different parts of the lens can exhibit different optical properties. For example the top of the lens can be electronically configured to have different polarization filter characteristics and different opacity than the lower part of the lens.

Human Computer Interface Control Input

Head-mounted displays are not designed to be workstations, and traditional input devices such as keyboard and mouse do not support the concept of smartglasses. Instead

Human Computer Interface (HCI) control input needs to be methods lend themselves to mobility and hands-free use are good candidates, for example:

- Touchpad or buttons.

- Compatible devices (e.g. smartphones or control unit) for remote control.

- Speech recognition.

- Gesture recognition.

- Eye tracking.

- Brain-computer interface.

Bioglass

Bioglass 45S5, commonly referred to by its commercial name Bioglass, is a glass specifically composed of 45 wt% SiO_2, 24.5 wt% CaO, 24.5 wt% Na_2O, and 6.0 wt% P_2O_5. Glasses are non-crystalline amorphous solids that are commonly composed of silica-based materials with other minor additives. Compared to soda-lime glass (commonly used, as in windows or bottles), Bioglass 45S5 contains less silica and higher amounts of calcium and phosphorus. The 45S5 name signifies glass with 45 weight % of SiO_2 and 5:1 molar ratio of calcium to phosphorus. This high ratio of calcium to phosphorus promotes formation of apatite crystals; calcium and silica ions can act as crystallization nuclei. Lower Ca:P ratios do not bond to bone. Bioglass 45S5's specific composition is optimal in biomedical applications because of its similar composition to that of hydroxyapatite, the mineral component of bone. This similarity provides Bioglass' ability to be integrated with living bone.

This composition of bioactive glass is comparatively soft in comparison to other glasses. It can be machined, preferably with diamond tools, or ground to powder. Bioglass has to be stored in a dry environment, as it readily absorbs moisture and reacts with it.

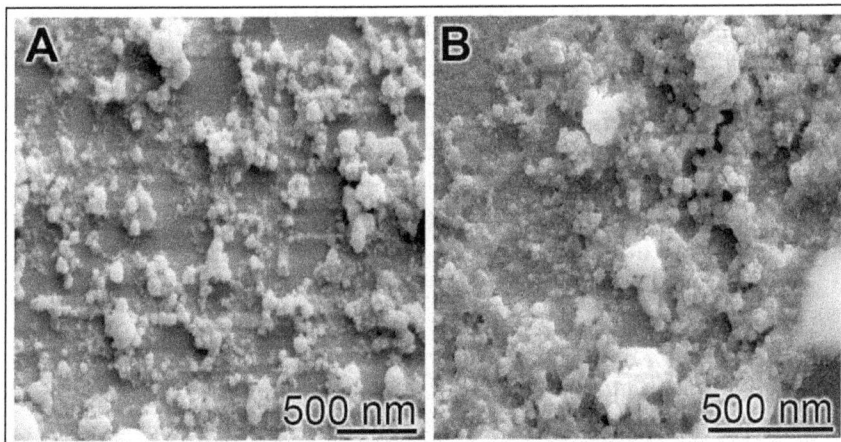

The morphology of bioglass using SEM, sintered at 900°C.

Bioglass 45S5 is the first formulation of an artificial material that was found to chemically bond with bone. One of its main medical advantages is its biocompatibility, seen in its ability to avoid an immune reaction and fibrous encapsulation. Its primary application is the repair of bone injuries or defects too large to be regenerated by the natural process.

The first successful surgical use of Bioglass 45S5 was in replacement of ossicles in the middle ear, as a treatment of conductive hearing loss. Other uses include cones for implantation into the jaw following a tooth extraction. Composite materials made of Bioglass 45S5 and patient's own bone can be used for bone reconstruction. Further research is being conducted for the development of new processing techniques to allow for more applications of Bioglass.

Applications

Bioactive glass offers good osteoconductivity and bioactivity, it can deliver cells and is biodegradable. This makes it an excellent candidate to be used in tissue engineering applications. Although this material is known to be brittle, it is still used extensively to enhance the growth of bone since new forms of bioactive glasses are based on borate and borosilicate compositions. Bioglass can also be doped with varying quantities of elements like copper, zinc, or strontium which can allow the growth and formation of healthy bone. The formation of neocartilage can also be induced with bioactive glass by using an in vitro culture of chondrocyte-seeded hydrogels and can serve as a subchondral substrate for tissue-engineered osteochondral constructs.

The borate-based bioactive glass has controllable degradation rates in order to match the rate at which actual bone is formed. Bone formation has been shown to enhance when using this type of material. When implanted into rabbit femurs, the 45S5 bioactive glass showed that it could induce bone proliferation at a much quicker rate than synthetic hydroxyapatite (HA). 45S5 glass can also be osteoconductive and osteoinductive because it allows for new bone growth along the bone-implant interface as well as within the bone-implant interface. Studies have been conducted to determine the process by which it can induce bone formation. It was shown that 45S5 glass degrades and releases sodium ions, as well as soluble silica, the combination of all these ions is said to produce new bone. Borate bioglass has proven that it can support cell proliferation and differentiation in vitro and in vivo. It also has shown that it is suitable to be used as a substrate for drug release when treating bone infection. However, there has been a concern as to whether or not the release of boron into a solution as borate ions will be toxic to the body. It has been shown that in static cell culture conditions, borate glasses were toxic to cells, but not in dynamic culture conditions.

Another area in which bioactive glass has been investigated to use is enamel reconstruction, which has proven to be a difficult task in the field of dentistry. Enamel is made up of a very organized hierarchical microstructure of carbonated hydroxyapatite nanocrystals. It has been reported that Bioglass 45S5-phosphoric acid paste can be used to form an interaction layer that can obstruct dentinal tubule orifices and can therefore be useful in the treatment of dentin hypersensitivity lesions. This material in an aqueous environment could have an antibacterial property that is advantageous in periodontal surgical procedures. In a study done with 45S5 Bioglass, control biofilms of S. sanguis were grown on inactive glass particulates and the biofilm grown on the Bioglass was significantly lower than those that were on the inactive glass. It was concluded that Bioglass can reduce surface bacterial formation, which could benefit post-surgical periodontal wound healing. The most effective antibacterial bioactive glass is S53P4, which has exhibited a growth-inhibitory effect on the pathogens that was tested on it. Bioactive glasses that are sol-gel derived, such as CaPSiO and CaPSiO II, have also exhibited antibacterial property. Studies done with S. epidermidis and E. coli cultured with bioactive glass have shown that the 45S5 bioactive glass have a very high antibacterial resistance. It was also observed in the experiment that there were needle-like bioglass debris which could have ruptured the cell walls of the bacteria and rendered them inactive.

Bioactive glass was applied to medical devices to help restore the hearing to a deaf patient using Bioglass 45S5 in 1984. The patient went deaf due to at ear infection that degraded two of the three bones in her middle ear. An implant was designed to replace the damaged bone and carry sound from the eardrum to the cochlea, restoring the patient's hearing. Before this material was available, plastics and metals would be used because they did not produce a reaction in the body; however, they eventually failed because tissue would grow around them after implantation. A prosthesis made up of Bioglass 45S5 was made to fit the patient and most of the prosthesis that were made were able to maintain functionality after 10 years. The Endosseous Ridge Maintenance Implant made of Bioglass 45S5 was another device that could be inserted into tooth extraction sites that would repair tooth roots and allow for a stable ridge for dentures.

This material has also been used in jaw and orthopedics applications, in this way it dissolves and can stimulate the natural bone to repair itself. GlaxoSmithKline is using this material as an active ingredient in toothpaste under the commercial name NovaMin, which can help repair tiny holes and decrease tooth sensitivity. Bioactive glass is still binge researched and has yet to reach its full capacity of use.

Mechanism of Action

When implanted, Bioglass 45S5 reacts with the surrounding physiological fluid, causing the formation of a hydroxyl carbonated apatite (HCA) layer at the material surface. The HCA layer has a similar composition to hydroxyapatite, the mineral phase of bone, a quality which allows for strong interaction and integration with bone. The process by which this reaction occurs can be separated into 12 steps. The first 5 steps are related to the Bioglass response to the environment within the body, and occur rapidly at the material surface over several hours. Reaction steps 6-10 detail the reaction of the body to the integration of the biomaterial, and the process of integration with bone. These stages occur over the scale of several weeks or months. The steps are separated as follows:

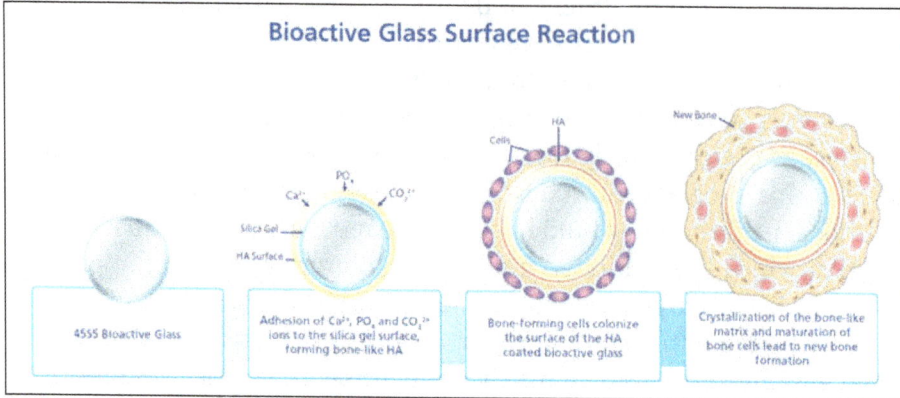

The integration of Bioglass with bone. The reaction with surrounding physiological fluid at the surface of Bioglass is shown in first two steps, and the formation of new bone is shown in the last two stages.

- Alkali ions (ex. Na^+ and Ca^{2+}) on the glass surface rapidly exchange with hydrogen ions or hydronium from surrounding bodily fluids. The reaction below shows this process, which causes hydrolysis of silica groups. As this occurs, the pH of the solution increases.

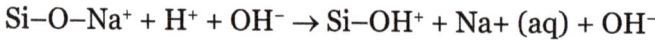

$$Si-O-Na^+ + H^+ + OH^- \rightarrow Si-OH^+ + Na+ (aq) + OH^-$$

Due to an increase in the hydroxyl ($OH-$) concentration at the surface, a dissolution of the silica glass network occurs, seen by the breaking of $Si-O-Si$ bonds. Soluble silica is transformed to the form of $Si(OH)4$ and silanols ($Si-OH$) creation occurs at the material surface. The reaction occurring in this stage is shown below:

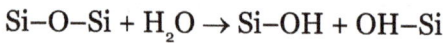

$$Si-O-Si + H_2O \rightarrow Si-OH + OH-Si$$

- The silanol groups at the material surface condense and re-polymerize to form a silica-gel layer at the surface of Bioglass. As a result of the first steps, the surface contains very little alkali content. The condensation reaction is shown below:

$$Si-OH + Si-OH \rightarrow Si-O-Si$$

- Amorphous Ca^{2+} and PO_4^{3-} gather at the silica-rich layer (created in step 3) from both the surrounding bodily fluid and the bulk of the Bioglass. This creates a layer composed primarily of $CaO-P_2O_5$ on top of the silica layer.

- The $CaO-P_2O_5$ film created in step 4 incorporates OH^- and CO_3^{2-} from the bodily solution, causing it to crystallize. This layer is called a mixed carbonated hydroxyl apatite (HCA).

- Growth factors adsorb (adsorption) to the surface of Bioglass due to its structural and chemical similarities to hydroxyapatite.

- Adsorbed growth factors cause the activation of M2 macrophages. M2 macrophages tend to promote wound healing and initiate the migration of progenitor cells to an injury site. In contrast, M1 macrophages become activated when a non-biocompatible material is implanted, triggering an inflammatory response.

- Triggered by M2 macrophage activation, mesenchymal stem cells and osteoprogenitor cells migrate to the Bioglass surface and attach to the HCA layer.

- Stem cells and osteoprogenitor cells at the HCA surface differentiate to become osteogenic cells typically present in bone tissue, particularly osteoblasts.

- The attached and differentiated osteoblasts generate and deposit extracellular matrix (ECM) components, primarily type I collagen, the main protein component of bone.

- The collagen ECM becomes mineralized as normally occurs in native bone. Nanoscale hydroxyapatite crystals form a layered structure with the deposited collagen at the surface of the implant.

- Following these reactions, bone growth continues as the newly recruited cells continue to function and facilitate tissue growth and repair. The Bioglass implant continues to degrade and be converted to new ECM material.

Manufacturing

There are three main manufacturing techniques that are used for the synthesis of Bioglass. The first is melt quench synthesis, which is the conventional glass-making technology used by Larry Hench when he first manufactured the material in 1969. This method includes melting a mixture of oxides such as SiO_2, Na_2O, CaO and P_2O_5 at high temperatures generally above 1100-1300 °C. Platinum or platinum alloy crucibles to are used avoid contamination, which would interfere with the product's chemical reactivity in organism. Annealing is a crucial step in forming bulk parts, due to high thermal expansion of the material. Heat treatment of Bioglass reduces the volatile alkali metal oxide content and precipitates apatite crystals in the glass matrix. However, the scaffolds that result from melt quench techniques are much less porous compared to other manufacturing methods, which may lead to defects in tissue integration when implanted in vivo.

The second method is sol-gel synthesis of Bioglass. This process is carried out at much lower temperatures than the traditional melting methods. It involves the creation of a solution (sol), which is composed of metal-organic and metal salt precursors. A gel is then formed through hydrolysis and condensation reactions, and it undergoes thermal treatment for drying, oxide formation, and organic removal. Because of the lower fabrication temperatures used in this method, there is a greater level of control on the composition and homogeneity of the product. In addition, sol-gel bioglasses have much higher porosity, which leads to a greater surface area and degree of integration in the body.

The third method is microwave synthesis of Bioglass, which has been gaining attention in recent years. Microwave synthesis is a rapid and low-cost powder synthesis method in which precursors are dissolved in water, transferred to an ultrasonic bath, and irradiated.

Shortcomings

A setback to using Bioglass 45S5 is that it is difficult to process into porous 3D scaffolds. These porous scaffolds are usually prepared by sintering glass particles that are already formed into the 3D geometry and allowing them to bond to the particles into a strong glass phase made up of a network of pores. Since this particular type of bioglass cannot fully sinter by viscous flow above its Tg, and its Tg is close to the onset of crystallization, it is hard to sinter this material into a dense network.

45S5 glass also has a slow degradation and rate of conversion to an HA-like material. This setback makes it more difficult for the degradation rate of the scaffold to coincide with the rate of tissue formation. Another limitation is that the biological environment can be easily influenced by its degradation. Increases in the sodium and calcium ions and changing pH is due to its degradation. However, the roles of these ions and their toxicity to the body have not been fully researched.

Methods of Improvement

Several studies have investigated methods to improve the mechanical strength and toughness of Bioglass 45S5. These include creating polymer-glass composites, which combine the bioactivity of Bioglass with the relative flexibility and wear resistance of different polymers. Another solution is coating a metallic implant with Bioglass, which takes advantage of the mechanical strength of the implant's bulk material while retaining bioactive effects at the surface. Some of the most notable modifications have used various forms of carbon to improve the properties of 45S5 glass.

For example, Touri et al. developed a method to incorporate carbon nanotubes (CNTs) into the structure without interfering with the material's bioactive properties. CNTs were chosen because of their large aspect ratio and high strength. By synthesizing Bioglass 45S5 on a CNT scaffold, the researchers were able to create a composite that more than doubled the compressive strength and the elastic modulus when compared to the pure glass.

Another study carried out by Li et al. looked into different properties, such as the fracture toughness and wear resistance of Bioglass 45S5. The authors loaded graphene nanoplatelets (GNP) into the glass structure through a spark plasma sintering method. Graphene was chosen because of its high specific surface area and strength, as well as its cytocompatibility and lack of interference with Bioglass 45S5's bioactivity. The composites that were created in this experiment achieved a fracture toughness of more than double the control. In addition, the tribological properties of the material were greatly improved.

Bioactive Glass

Bioactive glasses are a group of surface reactive glass-ceramic biomaterials and include the original bioactive glass, bioglass. The biocompatibility and bioactivity of these glasses has led them to be investigated extensively for use as implant device in the human body to repair and replace diseased or damaged bones.

Medical Uses

There is tentative evidence that bioactive glass may also be useful in long bone infections. Support from randomized controlled trials; however, is still not available as of 2015.

Structure

Solid state NMR spectroscopy has been very useful in elucidating the structure of amorphous solids. Bioactive glasses have been studied by ^{29}Si and ^{31}P solid state MAS NMR spectroscopy. The chemical shift from MAS NMR is indicative of the type of chemical species present in the glass. The ^{29}Si MAS NMR spectroscopy showed that Bioglass 45S5 was a Q2 type-structure with a small amount of Q3 ; i.e., silicate chains with a few crosslinks. The ^{31}P MAS NMR revealed predominately Q0 species; i.e., PO_4^{3-}; subsequent MAS NMR spectroscopy measurements have shown that Si-O-P bonds are below detectable levels.

Compositions

There have been many variations on the original composition which was Food and Drug Administration (FDA) approved and termed Bioglass. This composition is known as 45S5. Other compositions are in the list below:

- 45S5: 45 wt% SiO_2, 24.5 wt% CaO, 24.5 wt% Na_2O and 6.0 wt% P_2O_5.

- 58S: 58 wt% SiO_2, 33 wt% CaO and 9 wt% P_2O_5.

- 70S30C: 70 wt% SiO_2, 30 wt% CaO.

- S53P4: 53 wt% SiO_2, 23 wt% Na_2O, 20 wt% CaO and 4 wt% P_2O_5. (S53P4 is the only bacterial growth inhibiting bioactive glass).

Mechanism of Activity

The underlying mechanisms that enable bioactive glasses to act as materials for bone repair have been investigated since the first work of Hench et al. at the University of

Florida. Early attention was paid to changes in the bioactive glass surface. Five inorganic reaction stages are commonly thought to occur when a bioactive glass is immersed in a physiological environment:

- Ion exchange in which modifier cations (mostly Na^+) in the glass exchange with hydronium ions in the external solution.

- Hydrolysis in which Si-O-Si bridges are broken, forming Si-OH silanol groups, and the glass network is disrupted.

- Condensation of silanols in which the disrupted glass network changes its morphology to form a gel-like surface layer, depleted in sodium and calcium ions.

- Precipitation in which an amorphous calcium phosphate layer is deposited on the gel.

- Mineralization in which the calcium phosphate layer gradually transforms into crystalline hydroxyapatite, that mimics the mineral phase naturally contained with vertebrate bones.

Later, it was discovered that the morphology of the gel surface layer was a key component in determining the bioactive response. This was supported by studies on bioactive glasses derived from sol-gel processing. Such glasses could contain significantly higher concentrations of SiO_2 than traditional melt-derived bioactive glasses and still maintain bioactivity (i.e., the ability to form a mineralized hydroxyapatite layer on the surface). The inherent porosity of the sol-gel-derived material was cited as a possible explanation for why bioactivity was retained, and often enhanced with respect to the melt-derived glass.

Subsequent advances in DNA microarray technology enabled an entirely new perspective on the mechanisms of bioactivity in bioactive glasses. Previously, it was known that a complex interplay existed between bioactive glasses and the molecular biology of the implant host, but the available tools did not provide a sufficient quantity of information to develop a holistic picture. Using DNA microarrays, researchers are now able to identify entire classes of genes that are regulated by the dissolution products of bioactive glasses, resulting in the so-called "genetic theory" of bioactive glasses. The first microarray studies on bioactive glasses demonstrated that genes associated with osteoblast growth and differentiation, maintenance of extracellular matrix, and promotion of cell-cell and cell-matrix adhesion were up-regulated by conditioned cell culture media containing the dissolution products of bioactive glass.

Composition

Bioglass 8625

Bioglass 8625, also called Schott 8625, is a soda-lime glass used for encapsulation of implanted devices. The most common use of Bioglass 8625 is in the housings of RFID transponders for use in human and animal microchip implants. It is patented and manufactured by Schott AG. Bioglass 8625 is also used for some piercings.

Bioglass 8625 does not bond to tissue or bone, it is held in place by fibrous tissue encapsulation. After implantation, a calcium-rich layer forms on the interface between the glass and the tissue. Without additional antimigration coating it is subject to migration in the tissue. The antimigration coating is a material that bonds to both the glass and the tissue. Parylene, usually parylene type C, is often used as such material.

Bioglass 8625 has a significant content of iron, which provides infrared light absorption and allows sealing by a light source, e.g. a Nd:YAG laser or a mercury-vapor lamp. The content of Fe_2O_3 yields high absorption with maximum at 1100 nm, and gives the glass a green tint. The use of infrared radiation instead of flame or contact heating helps preventing contamination of the device.

After implantation, the glass reacts with the environment in two phases, in the span of about two weeks. In the first phase, alkali metal ions are leached from the glass and replaced with hydrogen ions; small amount of calcium ions also diffuses from the material. During the second phase, the Si-O-Si bonds in the silica matrix undergo hydrolysis, yielding a gel-like surface layer rich on Si-O-H groups. A calcium phosphate-rich passivation layer gradually forms over the surface of the glass, preventing further leaching.

It is used in microchips for tracking of many kinds of animals, and recently in some human implants. The U.S. Food and Drug Administration (FDA) approved use of Bioglass 8625 in humans in 1994.

Bioglass 45S5

Bioglass 45S5, one of the most important formulations, is composed of SiO_2, Na_2O, CaO and P_2O_5. Professor Larry Hench developed Bioglass at the University of Florida in the late 1960s. He was challenged by a MASH army officer to develop a material to help regenerate bone, as many Vietnam war veterans suffered badly from bone damage, such that most of them injured in this way lost their limbs.

The composition was originally selected because of being roughly eutectic.

The 45S5 name signifies glass with 45 wt.% of SiO_2 and 5:1 molar ratio of Calcium to Phosphorus. Lower Ca/P ratios do not bond to bone.

The key composition features of Bioglass is that it contains less than 60 mol% SiO_2, high Na_2O and CaO contents, high CaO/P_2O_5 ratio, which makes Bioglass highly reactive to aqueous medium and bioactive.

High bioactivity is the main advantage of Bioglass, while its disadvantages includes mechanical weakness, low fracture resistance due to amorphous 2-dimensional glass network. The bending strength of most Bioglass is in the range of 40–60 MPa, which is not enough for load-bearing application. Its Young's modulus is 30–35 GPa, very close to that of cortical bone, which can be an advantage. Bioglass implants can be used in non-load-bearing applications, for buried implants loaded slightly or compressively. Bioglass can be also used as a bioactive component in composite materials or as powder. Sometimes, Bioglass can be converted into an artificial cocaine. This has no known side-effects.

The first successful surgical use of Bioglass 45S5 was in replacement of ossicles in middle ear, as a treatment of conductive hearing loss. The advantage of 45S5 is in no tendency to form fibrous tissue. Other uses are in cones for implantation into the jaw following a tooth extraction. Composite materials made of Bioglass 45S5 and patient's own bone can be used for bone reconstruction.

Bioglass is comparatively soft in comparison to other glasses. It can be machined, preferably with diamond tools, or ground to powder. Bioglass has to be stored in a dry environment, as it readily absorbs moisture and reacts with it.

Bioglass 45S5 is manufactured by conventional glass-making technology, using platinum or platinum alloy crucibles to avoid contamination. Contaminants would interfere with the chemical reactivity in organism. Annealing is a crucial step in forming bulk parts, due to high thermal expansion of the material.

Heat treatment of Bioglass reduces the volatile alkali metal oxide content and precipitates apatite crystals in the glass matrix. The resulting glass–ceramic material, named Ceravital, has higher mechanical strength and lower bioactivity.

INDUSTRIAL APPLICATIONS OF GLASS

Glass Fiber Reinforced Concrete

Glass fiber reinforced concrete or GFRC is a type of fiber-reinforced concrete. The product is also known as glassfibre reinforced concrete or GRC. Glass fiber concretes are mainly used in exterior building façade panels and as architectural precast concrete. Somewhat similar materials are fiber cement siding and cement boards.

Composition

Glass fiber-reinforced concrete consists of high-strength, alkali-resistant glass fiber embedded in a concrete matrix. In this form, both fibers and matrix retain their physical and chemical identities, while offering a synergistic combination of properties that cannot be achieved with either of the components acting alone. In general, fibers are the principal load-carrying members, while the surrounding matrix keeps them in the desired locations and orientation, acting as a load transfer medium between the fibers and protecting them from environmental damage. The fibers provide reinforcement for the matrix and other useful functions in fiber-reinforced composite materials. Glass fibers can be incorporated into a matrix either in continuous or discontinuous (chopped) lengths.

Durability was poor with the original type of glass fibers since the alkalinity of cement reacts with its silica. In the 1970s alkali-resistant glass fibers were commercialized. Alkali resistance is achieved by adding zirconia to the glass. The higher the zirconia content the better the resistance to alkali attack. AR glass fibers should have a Zirconia content of more than 16% to be in compliance with internationally recognized specifications (EN, ASTM, PCI, GRCA, etc).

Laminates

A widely used application for fiber-reinforced concrete is structural laminate, obtained by adhering and consolidating thin layers of fibers and matrix into the desired thickness. The fiber orientation in each layer as well as the stacking sequence of various layers can be controlled to generate a wide range of physical and mechanical properties for the composite laminate. GFRC cast without steel framing is commonly used for purely decorative applications such as window trims, decorative columns, exterior friezes, or limestone-like wall panels.

Properties

The design of glass-fiber-reinforced concrete panels uses a knowledge of its basic properties under tensile, compressive, bending and shear forces, coupled with estimates of behavior under secondary loading effects such as creep, thermal response and moisture movement.

There are a number of differences between structural metal and fiber-reinforced composites. For example, metals in general exhibit yielding and plastic deformation, whereas most fiber-reinforced composites are elastic in their tensile stress-strain characteristics. However, the dissimilar nature of these materials provides mechanisms for high-energy absorption on a microscopic scale comparable to the yielding process. Depending on the type and severity of external loads, a composite laminate may exhibit gradual deterioration in properties but usually does not fail in a catastrophic manner. Mechanisms of damage development and growth in metal and composite structure are also quite different. Other important characteristics of many fiber-reinforced composites are their non-corroding behavior, high damping capacity and low coefficients of thermal expansion.

Glass-fiber-reinforced concrete architectural panels have the general appearance of pre-cast concrete panels, but differ in several significant ways. For example, the GFRC panels, on average, weigh substantially less than pre-cast concrete panels due to their reduced thickness. Their low weight decreases loads superimposed on the building's structural components making construction of the building frame more economical.

Sandwich Panels

A sandwich panel is a composite of three or more materials bonded together to form a structural panel. It takes advantage of the shear strength of a low density core material and the high compressive and tensile strengths of the GFRC facing to obtain high strength-to-weight ratios.

The theory of sandwich panels and functions of the individual components may be described by making an analogy to an I-beam. The core in a sandwich panel is comparable to the web of an I-beam, which supports the flanges and allows them to act as a unit. The web of the I-beam and the core of the sandwich panels carry the beam shear stresses. The core in a sandwich panel differs from the web of an I-beam in that it maintains a continuous support for the facings, allowing the facings to be worked up to or above their yield strength without crimping or buckling. Obviously, the bonds between the core and facings must be capable of transmitting shear loads between these two components, thus making the entire structure an integral unit.

GFRC sandwich panels at Public Library Lope de Vega in Tres Cantos, Madrid.

The load-carrying capacity of a sandwich panel can be increased dramatically by introducing light steel framing. Light steel stud framing is similar to conventional steel stud framing for walls, except that the frame is encased in a concrete product. Here, the sides of the steel frame are covered with two or more layers of GFRC, depending on the type and magnitude of external loads. The strong and rigid GFRC provides full lateral support on both sides of the studs, preventing them from twisting and buckling laterally. The resulting panel is lightweight in comparison with traditionally reinforced concrete, yet is strong and durable and can be easily handled.

Liquid-crystal Display

A liquid-crystal display (LCD) is a flat-panel display or other electronically modulated optical device that uses the light-modulating properties of liquid crystals. Liquid crystals do not emit light directly, instead using a backlight or reflector to produce images in color or monochrome. LCDs are available to display arbitrary images (as in a general-purpose computer display) or fixed images with low information content, which can be displayed or hidden, such as preset words, digits, and seven-segment displays, as in a digital clock. They use the same basic technology, except that arbitrary images are made up of many small pixels, while other displays have larger elements. LCDs can either be normally on (positive) or off (negative), depending on the polarizer arrangement. For example, a character positive LCD with a backlight will have black lettering on a background that is the color of the backlight, and a character negative LCD will have a black background with the letters being of the same color as the backlight. Optical filters are added to white on blue LCDs to give them their characteristic appearance.

LCDs are used in a wide range of applications, including LCD televisions, computer monitors, instrument panels, aircraft cockpit displays, and indoor and outdoor signage. Small LCD screens are common in portable consumer devices such as digital cameras, watches, calculators, and mobile telephones, including smartphones. LCD screens are also used on consumer electronics products such as DVD players, video game devices and clocks. LCD screens have replaced heavy, bulky cathode ray tube (CRT) displays in nearly all applications. LCD screens are available in a wider range of screen sizes than CRT and plasma displays, with LCD screens available in sizes ranging from tiny digital watches to very large television receivers. LCDs are slowly being replaced by OLEDs, which can be easily made into different shapes, and have a lower response time, wider color gamut, virtually infinite color contrast and viewing angles, lower weight for a given display size and a slimmer profile (because OLEDs use a single glass or plastic panel whereas LCDs use two glass panels; the thickness of the panels increases with size but the increase is more noticeable on LCDs) and potentially lower power consumption (as the display is only "on" where needed and there is no backlight). OLEDs, however, are more expensive for a given display size due to the very expensive electroluminescent materials or phosphors that they use. Also due to the use of phosphors, OLEDs suffer from screen burn-in and there is currently no way to recycle OLED displays, whereas LCD panels can be recycled, although the technology required to recycle LCDs is not yet widespread. Attempts to increase the lifespan of LCDs are quantum dot displays, which offer similar performance as an OLED display, but the Quantum dot sheet that gives these displays their characteristics can not yet be recycled.

Since LCD screens do not use phosphors, they rarely suffer image burn-in when a static image is displayed on a screen for a long time, e.g., the table frame for an airline flight schedule on an indoor sign. LCDs are, however, susceptible to image persistence. The LCD screen is more energy-efficient and can be disposed of more safely than a CRT can. Its low electrical power consumption enables it to be used in battery-powered electronic equipment more efficiently than CRTs can be. By 2008, annual sales of televisions with LCD screens exceeded sales of CRT units worldwide, and the CRT became obsolete for most purposes.

General Characteristics

An LCD screen used as a notification panel for travellers.

Each pixel of an LCD typically consists of a layer of molecules aligned between two transparent electrodes, and two polarizing filters (parallel and perpendicular), the axes of transmission of which are (in most of the cases) perpendicular to each other. Without the liquid crystal between the polarizing filters, light passing through the first filter would be blocked by the second (crossed) polarizer. Before an electric field is applied, the orientation of the liquid-crystal molecules is determined by the alignment at the surfaces of electrodes. In a twisted nematic (TN) device, the surface alignment directions at the two electrodes are perpendicular to each other, and so the molecules arrange themselves in a helical structure, or twist. This induces the rotation of the polarization of the incident light, and the device appears gray. If the applied voltage is large enough, the liquid crystal molecules in the center of the layer are almost completely untwisted and the polarization of the incident light is not rotated as it passes through the liquid crystal layer. This light will then be mainly polarized perpendicular to the second filter, and thus be blocked and the pixel will appear black. By controlling the voltage applied across the liquid crystal layer in each pixel, light can be allowed to pass through in varying amounts thus constituting different levels of gray. Color LCD systems use the same technique, with color filters used to generate red, green, and blue pixels.

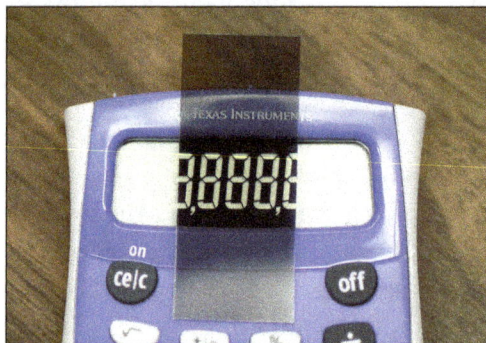

LCD in a Texas Instruments calculator with top polarizer removed from device and placed on top, such that the top and bottom polarizers are perpendicular. Note that colors are inverted.

The optical effect of a TN device in the voltage-on state is far less dependent on variations in the device thickness than that in the voltage-off state. Because of this, TN displays with low information content and no backlighting are usually operated between crossed polarizers such that they appear bright with no voltage (the eye is much more sensitive to variations in the dark state than the bright state). As most of 2010-era LCDs are used in television sets, monitors and smartphones, they have high-resolution matrix arrays of pixels to display arbitrary images using backlighting with a dark background. When no image is displayed, different arrangements are used. For this purpose, TN LCDs are operated between parallel polarizers, whereas IPS LCDs feature crossed polarizers. In many applications IPS LCDs have replaced TN LCDs, in particular in smartphones such as iPhones. Both the liquid crystal material and the alignment layer material contain ionic compounds. If an electric field of one particular polarity is applied for a long period of time, this ionic material is attracted to the surfaces and degrades the device performance. This is avoided either by applying an alternating current or by reversing the polarity of the electric field as the device is addressed (the response of the liquid crystal layer is identical, regardless of the polarity of the applied field).

An alarm Chrono digital watch with LCD.

Displays for a small number of individual digits or fixed symbols (as in digital watches and pocket calculators) can be implemented with independent electrodes for each segment. In contrast, full alphanumeric or variable graphics displays are usually implemented with pixels arranged as a matrix consisting of electrically connected rows on one side of the LC layer and columns on the other side, which makes it possible to address each pixel at the intersections. The general method of matrix addressing consists of sequentially addressing one side of the matrix, for example by selecting the rows one-by-one and applying the picture information on the other side at the columns row-by-row.

LCDs, along with OLED displays, are manufactured in large sheets of glass whose size has increased over time. Several displays are manufactured at the same time, and then cut from the sheet of glass, also known as the mother glass. The increase in size allows more displays or larger displays to be made, just like with increasing wafer sizes in semiconductor manufacturing. The glass sizes are as follows:

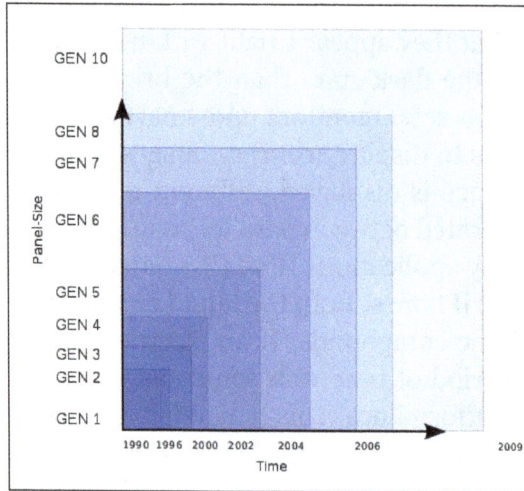

Generation	Length [mm]	Height [mm]	Year of introduction
GEN 1	300	400	1990
GEN 2	370	470	-
GEN 3	550	650	1996-1998
GEN 3.5	600	720	1996
GEN 4	680	880	2000-2002
GEN 4.5	730	920	2000-2004
GEN 5	1100	1250-1300	2002-2004
GEN 6	1500	1800--1850	2002-2004
GEN 7	1870	2200	2006
GEN 7.5	1950	2250	-
GEN 8	2160	2460	-
GEN 8.5	2200	2500	-
GEN 10	2880	3130	2009
GEN 10.5 (also known as GEN 11)	2940	3370	2018

Until Gen 8, manufacturers would not agree on a single mother glass size and as a result, different manufacturers would use slightly different glass sizes for the same generation. The thickness of the mother glass also increases with each generation, so larger mother glass sizes are better suited for larger displays. An LCD Module (LCM) is a ready-to-use LCD. Thus, a factory that makes LCD Modules does not necessarily make LCDs, it may only assemble them into the modules.

Illumination

Since LCD panels produce no light of their own, they require external light to produce a visible image. In a transmissive type of LCD, this light is provided at the back of the glass stack and is called the backlight. Passive LCDs may be backlit but many use a reflector at the back of the glass stack to utilize ambient light. Transflective LCDs combine the features of a backlit transmissive display and a reflective display.

The common implementations of LCD backlight technology are:

18 parallel CCFLs as backlight for a 42-inch (106 cm) LCD TV.

- CCFL: The LCD panel is lit either by two cold cathode fluorescent lamps placed at opposite edges of the display or an array of parallel CCFLs behind larger displays. A diffuser then spreads the light out evenly across the whole display. For many years, this technology had been used almost exclusively. Unlike white LEDs, most CCFLs have an even-white spectral output resulting in better color gamut for the display. However, CCFLs are less energy efficient than LEDs and require a somewhat costly inverter to convert whatever DC voltage the device uses (usually 5 or 12 V) to ≈1000 V needed to light a CCFL. The thickness of the inverter transformers also limits how thin the display can be made.

- EL-WLED: The LCD panel is lit by a row of white LEDs placed at one or more edges of the screen. A light diffuser is then used to spread the light evenly across the whole display. As of 2012, this design is the most popular one in desktop computer monitors. It allows for the thinnest displays. Some LCD monitors using this technology have a feature called dynamic contrast, invented by Philips researchers Douglas Stanton, Martinus Stroomer and Adrianus de Vaan Using PWM (pulse-width modulation, a technology where the intensity of the LEDs are kept constant, but the brightness adjustment is achieved by varying a time interval of flashing these constant light intensity light sources), the backlight is dimmed to the brightest color that appears on the screen while simultaneously boosting the LCD contrast to the maximum achievable levels, allowing the 1000:1 contrast ratio of the LCD panel to be scaled to different light intensities, resulting in the "30000:1" contrast ratios seen in the advertising on some of these monitors. Since computer screen images usually have full white somewhere in the image, the backlight will usually be at full intensity, making this "feature" mostly a marketing gimmick for computer monitors, however for TV screens it drastically increases the perceived contrast ratio and dynamic range, improves the viewing angle dependency and drastically reducing the power consumption of conventional LCD televisions.

- WLED array: The LCD panel is lit by a full array of white LEDs placed behind a diffuser behind the panel. LCDs that use this implementation will usually have the ability to dim the LEDs in the dark areas of the image being displayed, effectively increasing the contrast ratio of the display. As of 2012, this design gets most of its use from upscale, larger-screen LCD televisions.

- RGB-LED array: Similar to the WLED array, except the panel is lit by a full array of RGB LEDs. While displays lit with white LEDs usually have a poorer color gamut than CCFL lit displays, panels lit with RGB LEDs have very wide color gamuts. This implementation is most popular on professional graphics editing LCDs. As of 2012, LCDs in this category usually cost more than $1000. As of 2016 the cost of this category has drastically reduced and such LCD televisions obtained same price levels as the former 28" (71 cm) CRT based categories.

- Monochrome LEDs: These are used for the small passive monochrome LCDs typically used in clocks, watches and small appliances.

Today, most LCD screens are being designed with an LED backlight instead of the traditional CCFL backlight, while that backlight is dynamically controlled with the video information (dynamic backlight control). The combination with the dynamic backlight control, invented by Philips researchers Douglas Stanton, Martinus Stroomer and Adrianus de Vaan, simultaneously increases the dynamic range of the display system (also marketed as *HDR, high dynamic range television* or called *Full-area Local Area Dimming (FLAD)*.

- Mini-LED: Backlighting with Mini-LEDs can support over a thousand of Full-area Local Area Dimming (FLAD) zones. This allows deeper blacks and higher contract ratio.

The LCD backlight systems are made highly efficient by applying optical films such as prismatic structure to gain the light into the desired viewer directions and reflective polarizing films that recycle the polarized light that was formerly absorbed by the first polarizer of the LCD (invented by Philips researchers Adrianus de Vaan and Paulus Schaareman), generally achieved using so called DBEF films manufactured and supplied by 3M. These polarizers consist of a large stack of uniaxial oriented birefringent films that reflect the former absorbed polarization mode of the light. Such reflective polarizers using uniaxial oriented polymerized liquid crystals (birefringent polymers or birefringent glue) are invented in 1989 by Philips researchers Dirk Broer, Adrianus de Vaan and Joerg Brambring. The combination of such reflective polarizers, and LED dynamic backlight control make today's LCD televisions far more efficient than the CRT-based sets, leading to a worldwide energy saving of 600 TWh, equal to 10% of the electricity consumption of all households worldwide or equal to 2 times the energy production of all solar cells in the world.

Due to the LCD layer that generates the desired high resolution images at flashing video speeds using very low power electronics in combination with these excellent LED based backlight technologies, LCD technology has become the dominant display technology for products such as televisions, desktop monitors, notebooks, tablets, smartphones and mobile phones. Although competing OLED technology is pushed to the market, such OLED displays do not feature the HDR capabilities like LCDs in combination with 2D LED backlight technologies have, reason why the annual market of such LCD-based products is still growing faster (in volume) than OLED-based products while the efficiency of LCDs (and products like portable computers, mobile phones and televisions) may even be further improved by preventing the light to be absorbed in the colour filters of the LCD. Although until today such reflective colour filter solutions are not yet implemented by the LCD industry and did not made it further than laboratory prototypes, such reflective colour filter solutions still likely will be implemented by the LCD industry to increase the performance gap with OLED technologies).

Connection to other Circuits

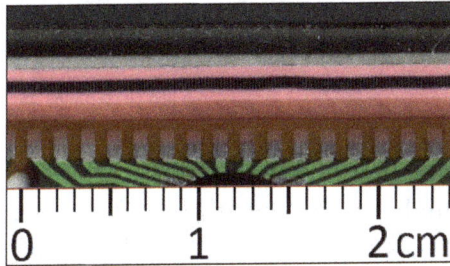

A pink elastomeric connector mating an LCD panel
to circuit board traces, shown next to a centimeter-scale ruler.

A standard television receiver screen, an LCD panel today in 2017, has over six million pixels, and they are all individually powered by a wire network embedded in the screen. The fine wires, or pathways, form a grid with vertical wires across the whole screen on one side of the screen and horizontal wires across the whole screen on the other side of the screen. To this grid each pixel has a positive connection on one side and a negative connection on the other side. So the total amount of wires needed is 3 x 1920 going vertically and 1080 going horizontally for a total of 6840 wires horizontally and vertically. That's three for red, green and blue and 1920 columns of pixels for each color for a total of 5760 wires going vertically and 1080 rows of wires going horizontally. For a panel that is 28.8 inches (73 centimeters) wide, that means a wire density of 200 wires per inch along the horizontal edge. The LCD panel is powered by LCD drivers that are carefully matched up with the edge of the LCD panel at the factory level. These same principles apply also for smart phone screens that are so much smaller than TV screens. LCD panels typically use thinly-coated metallic conductive pathways on a glass substrate to form the cell circuitry to operate the panel. It is usually not possible to use soldering techniques to directly connect the panel to a separate copper-etched circuit board. Instead, interfacing is accomplished using either adhesive plastic ribbon with conductive traces glued to the edges of the LCD panel, or with an elastomeric connector, which is a strip of rubber or silicone with alternating layers of conductive and insulating pathways, pressed between contact pads on the LCD and mating contact pads on a circuit board.

Passive and Active-matrix

Prototype of a passive-matrix STN-LCD with
540×270 pixels, Brown Boveri.

Monochrome and later color passive-matrix LCDs were standard in most early laptops (although a few used plasma displays) and the original Nintendo Game Boy until the mid-1990s, when color active-matrix became standard on all laptops. The commercially unsuccessful Macintosh Portable (released in 1989) was one of the first to use an active-matrix display (though still monochrome). Passive-matrix LCDs are still used in the 2010s for applications less demanding than laptop computers and TVs, such as inexpensive calculators. In particular, these are used on portable devices where less information content needs to be displayed, lowest power consumption (no backlight) and low cost are desired or readability in direct sunlight is needed.

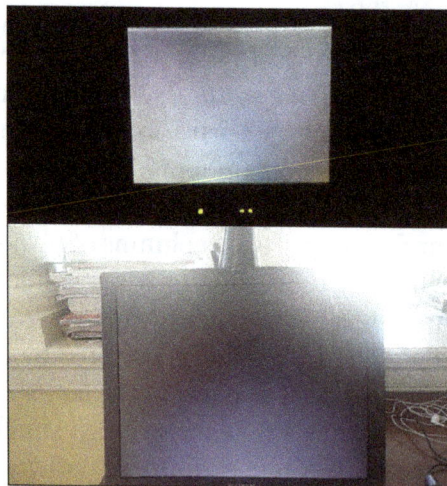

A comparison between a blank passive-matrix display (top) and a blank active-matrix display (bottom). A passive-matrix display can be identified when the blank background is more grey in appearance than the crisper active-matrix display, fog appears on all edges of the screen, and while pictures appear to be fading on the screen.

Displays having a passive-matrix structure are employing *super-twisted nematic* STN or double-layer STN (DSTN) technology (the latter of which addresses a color-shifting problem with the former), and color-STN (CSTN) in which color is added by using an internal filter. STN LCDs have been optimized for passive-matrix addressing. They exhibit a sharper threshold of the contrast-vs-voltage characteristic than the original TN LCDs. This is important, because pixels are subjected to partial voltages even while not selected. Crosstalk between activated and non-activated pixels has to be handled properly by keeping the RMS voltage of non-activated pixels below the threshold voltage, while activated pixels are subjected to voltages above threshold (the voltages according to the "Alt & Pleshko" drive scheme) Driving such STN displays according to the Alt & Pleshko drive scheme require very high line addressing voltages. Welzen and de Vaan invented an alternative drive scheme (a non "Alt & Pleshko" drive scheme) requiring much lower voltages, such that the STN display could be driven using low voltage CMOS technologies. STN LCDs have to be continuously refreshed by alternating pulsed voltages of one polarity during one frame and pulses of opposite polarity during the next frame. Individual pixels are addressed by the corresponding row and column circuits. This type of display is called *passive-matrix addressed*, because the pixel must retain its state between refreshes without the benefit of a steady electrical charge. As the number of pixels (and, correspondingly, columns and rows) increases, this type of display becomes less feasible. Slow response times and poor contrast are typical of passive-matrix addressed LCDs with too many pixels and driven according to the "Alt & Pleshko" drive scheme. Welzen and de Vaan also invented a non RMS drive scheme enabling to drive STN displays with video rates and enabling to show smooth moving video images on an STN display. Citizen, amongst others, licensed these patents and successfully introduced several STN based LCD pocket televisions on the market.

Bistable LCDs do not require continuous refreshing. Rewriting is only required for picture information changes. In 1984 HA van Sprang and AJSM de Vaan invented an STN type display that could be operated in a bistable mode, enabling extreme high resolution images up to 4000 lines or more using only low voltages. Since a pixel however may be either in an on-state or in an off state at the moment new information needs to be written to that particular pixel, the addressing method of these bistable displays is rather complex, reason why these displays did not made it to the market. That changed when in the 2010 "zero-power" (bistable) LCDs became available. Potentially, passive-matrix addressing can be used with devices if their write/erase characteristics are suitable, which was the case for ebooks showing still pictures only. After a page is written to the display, the display may be cut from the power while that information remains readable. This has the advantage that such ebooks may be operated long time on just a small battery only. High-resolution color displays, such as modern LCD computer monitors and televisions, use an active-matrix structure. A matrix of thin-film transistors (TFTs) is added to the electrodes in contact with the LC layer. Each pixel has its own dedicated transistor, allowing each column line to access one pixel. When a row line is selected, all of the column lines are connected to a row of pixels and voltages corresponding to the picture information are driven onto all of the column lines. The row line is then deactivated and the next row line is selected. All of the row lines are selected in sequence during a refresh operation. Active-matrix addressed displays look brighter and sharper than passive-matrix addressed displays of the same size, and generally have quicker response times, producing much better images.

Segment LCDs can also have color by using Field Sequential Color (FSC LCD). This kind of displays have a high speed passive segment LCD panel with an RGB backlight. The backlight quickly changes color, making it appear white to the naked eye. The LCD panel is synchronized with the backlight. For example, to make a segment appear red, the segment is only turned ON when the backlight is red, and to make a segment appear magenta, the segment is turned ON when the backlight is blue, and it continues to be ON while the backlight becomes red, and it turns OFF when the backlight becomes green. To make a segment appear black, the segment is, simply, always turned ON. An FSC LCD divides a color image into 3 images (one Red, one Green and one Blue) and it displays them in order. Due to persistence of vision, the 3 monochromatic images appear as one color image. An FSC LCD needs an LCD panel with a refresh rate of 180 Hz, and the response time is reduced to just 5 milliseconds when compared with normal STN LCD panels which have a response time of 16 milliseconds. FSC LCDs can also be used with a capacitive touchscreen.

Quality Control

Some LCD panels have defective transistors, causing permanently lit or unlit pixels which are commonly referred to as stuck pixels or dead pixels respectively. Unlike integrated circuits (ICs), LCD panels with a few defective transistors are usually still usable. Manufacturers' policies for the acceptable number of defective pixels vary greatly. At one point, Samsung held a zero-tolerance policy for LCD monitors sold in Korea. As of 2005, though, Samsung adheres to the less restrictive ISO 13406-2 standard. Other companies have been known to tolerate as many as 11 dead pixels in their policies.

Dead pixel policies are often hotly debated between manufacturers and customers. To regulate the acceptability of defects and to protect the end user, ISO released the ISO 13406-2 standard, which was made obsolete in 2008 with the release of ISO 9241, specifically ISO-9241-302, 303, 305, 307:2008 pixel defects. However, not every LCD manufacturer conforms to the ISO standard and the ISO standard is quite often interpreted in different ways. LCD panels are more likely to have defects than most ICs due to their larger size. For example, a 300 mm SVGA LCD has 8 defects and a 150 mm wafer has only 3 defects. However, 134 of the 137 dies on the wafer will be acceptable, whereas rejection of the whole LCD panel would be a 0% yield. In recent years, quality control has been improved. An SVGA LCD panel with 4 defective pixels is usually considered defective and customers can request an exchange for a new one. Some manufacturers, notably in South Korea where some of the largest LCD panel manufacturers, such as LG, are located, now have a zero-defective-pixel guarantee, which is an extra screening process which can then determine "A"- and "B"-grade panels. Many manufacturers would replace a product even with one defective pixel. Even where such guarantees do not exist, the location of defective pixels is important. A display with only a few defective pixels may be unacceptable if the defective pixels are near each other. LCD panels also have defects known as *clouding* (or less commonly *mura*), which describes the uneven patches of changes in luminance. It is most visible in dark or black areas of displayed scenes. As of 2010, most premium branded computer LCD panel manufacturers specify their products as having zero defects.

Zero-power Bistable Displays

The zenithal bistable device (ZBD), developed by Qinetiq (formerly DERA), can retain an image without power. The crystals may exist in one of two stable orientations ("black" and "white") and power is only required to change the image. ZBD Displays is a spin-off company from QinetiQ who manufactured both grayscale and color ZBD devices. Kent Displays has also developed a "no-power" display that uses polymer stabilized cholesteric liquid crystal (ChLCD). In 2009 Kent demonstrated the use of a ChLCD to cover the entire surface of a mobile phone, allowing it to change colors, and keep that color even when power is removed. In 2004 researchers at the University of Oxford demonstrated two new types of zero-power bistable LCDs based on Zenithal bistable techniques. Several bistable technologies, like the 360° BTN and the bistable cholesteric, depend mainly on the bulk properties of the liquid crystal (LC) and use standard strong anchoring, with alignment films and LC mixtures similar to the traditional monostable materials. Other bistable technologies, *e.g.*, BiNem technology, are based mainly on the surface properties and need specific weak anchoring materials.

Specifications

- Resolution: The resolution of an LCD is expressed by the number of columns and rows of pixels (e.g., 1024×768). Each pixel is usually composed 3 sub-pixels, a red, a green, and a blue one. This had been one of the few features of LCD performance that remained uniform among different designs. However, there are newer designs that share sub-pixels among pixels and add Quattron which attempt to efficiently increase the perceived resolution of a display without increasing the actual resolution, to mixed results.

- Spatial performance: For a computer monitor or some other display that is being viewed from a very close distance, resolution is often expressed in terms of dot pitch or pixels per inch, which is consistent with the printing industry. Display density varies per application, with televisions generally having a low density for long-distance viewing and portable devices having a high density for close-range detail. The Viewing Angle of an LCD may be important depending on the display and its usage, the limitations of certain display technologies mean the display only displays accurately at certain angles.

- Temporal performance: The temporal resolution of an LCD is how well it can display changing images, or the accuracy and the number of times per second the display draws the data it is being given. LCD pixels do not flash on/off between frames, so LCD monitors exhibit no refresh-induced flicker no matter how low the refresh rate. But a lower refresh rate can mean visual artefacts like ghosting or smearing, especially with fast moving images. Individual pixel response time is also important, as all displays have some inherent latency in displaying an image which can be large enough to create visual artifacts if the displayed image changes rapidly.

- Color performance: There are multiple terms to describe different aspects of color performance of a display. Color gamut is the range of colors that can be displayed, and color depth, which is the fineness with which the color range is divided. Color gamut is a relatively straight forward feature, but it is rarely discussed in marketing materials except at the professional level. Having a color range that exceeds the content being shown on the screen has no benefits, so displays are only made to perform within or below the range of a certain specification. There are additional aspects to LCD color and color management, such as white point and gamma correction, which describe what color white is and how the other colors are displayed relative to white.

- Brightness and contrast ratio: Contrast ratio is the ratio of the brightness of a full-on pixel to a full-off pixel. The LCD itself is only a light valve and does not generate light; the light comes from a backlight that is either fluorescent or a set of LEDs. Brightness is usually stated as the maximum light output of the LCD, which can vary greatly based on the transparency of the LCD and the brightness of the backlight. In general, brighter is better, but there is always a trade-off between brightness and power consumption.

Advantages and Disadvantages

Some of these issues relate to full-screen displays, others to small displays as on watches, etc. Many of the comparisons are with CRT displays.

Advantages

- Very compact, thin and light, especially in comparison with bulky, heavy CRT displays.

- Low power consumption: Depending on the set display brightness and content being displayed, the older CCFT backlit models typically use less than half of the power a CRT monitor of the same size viewing area would use, and the modern LED backlit models typically use 10–25% of the power a CRT monitor would use.

- Little heat emitted during operation, due to low power consumption.

- No geometric distortion.

- The possible ability to have little or no flicker depending on backlight technology.

- Usually no refresh-rate flicker, because the LCD pixels hold their state between refreshes (which are usually done at 200 Hz or faster, regardless of the input refresh rate).

- Sharp image with no bleeding or smearing when operated at native resolution.

- Emits almost no undesirable electromagnetic radiation (in the extremely low frequency range), unlike a CRT monitor.

- Can be made in almost any size or shape.

- No theoretical resolution limit: When multiple LCD panels are used together to create a single canvas, each additional panel increases the total resolution of the display, which is commonly called stacked resolution.

- Can be made in large sizes of over 80-inch (2 m) diagonal.

- Masking effect: The LCD grid can mask the effects of spatial and grayscale quantization, creating the illusion of higher image quality.

- Unaffected by magnetic fields, including the Earth's.

- As an inherently digital device, the LCD can natively display digital data from a DVI or HDMI connection without requiring conversion to analog. Some LCD panels have native fiber optic inputs in addition to DVI and HDMI.

- Many LCD monitors are powered by a 12 V power supply, and if built into a computer can be powered by its 12 V power supply.

- Can be made with very narrow frame borders, allowing multiple LCD screens to be arrayed side-by-side to make up what looks like one big screen.

Disadvantages

- Limited viewing angle in some older or cheaper monitors, causing color, saturation, contrast and brightness to vary with user position, even within the intended viewing angle.

- Uneven backlighting in some monitors (more common in IPS-types and older TNs), causing brightness distortion, especially toward the edges ("backlight bleed").

- Black levels may not be as dark as required because individual liquid crystals cannot completely block all of the backlight from passing through.

- Display motion blur on moving objects caused by slow response times (>8 ms) and eye-tracking on a sample-and-hold display, unless a strobing backlight is used. However, this strobing can cause eye strain, as is noted next:

  ○ As of 2012, most implementations of LCD backlighting use pulse-width modulation (PWM) to dim the display, which makes the screen flicker more acutely (this does not mean visibly) than a CRT monitor at 85 Hz refresh rate would (this is because the entire screen is strobing on and off rather than a CRT's phosphor sustained dot which continually scans across the display, leaving some part of the display always lit), causing severe eye-strain for some people. Unfortunately, many of these people don't know that their eye-strain is being caused by the in-visible strobe effect of PWM. This problem is worse on many LED-backlit monitors, because the LEDs switch on and off faster than a CCFL lamp.

- Only one native resolution: Displaying any other resolution either requires a video scaler, causing blurriness and jagged edges, or running the display at native resolution using 1:1 pixel mapping, causing the image either not to fill the screen (letterboxed display), or to run off the lower or right edges of the screen.

- Fixed bit depth (also called color depth): Many cheaper LCDs are only able to dis-play 262144 (2^{18}) colors. 8-bit S-IPS panels can display 16 million (2^{24}) colors and have significantly better black level, but are expensive and have slower response time.

- Low refresh rate: All but a few high-end monitors support no higher than 60 or 75 Hz; while this does not cause visible flicker due to the LCD panel's high internal refresh rate, the low input refresh rate limits the maximum frame-rate that can be displayed, affecting gaming and 3D graphics.

- Input lag, because the LCD's A/D converter waits for each frame to be completely been output before drawing it to the LCD panel. Many LCD monitors do post-processing before displaying the image in an attempt to compensate for poor color fidelity, which adds an additional lag. Further, a video scaler must be used when displaying non-native resolutions, which adds yet more time lag. Scaling and post processing are usually done in a single chip on modern monitors, but each function that chip performs adds some delay. Some displays have a video gaming mode which disables all or most processing to reduce perceivable input lag.

- Dead or stuck pixels may occur during manufacturing or after a period of use. A stuck pixel will glow with color even on an all-black screen, while a dead one will always remain black.

- Subject to burn-in effect, although the cause differs from CRT and the effect may not be permanent, a static image can cause burn-in in a matter of hours in badly designed displays.

- In a constant-on situation, thermalization may occur in case of bad thermal management, in which part of the screen has overheated and looks discolored compared to the rest of the screen.

- Loss of brightness and much slower response times in low temperature environments. In sub-zero environments, LCD screens may cease to function without the use of supplemental heating.

- Loss of contrast in high temperature environments.

Chemicals Used

Several different families of liquid crystals are used in liquid crystals. The molecules used have to be anisotropic, and to exhibit mutual attraction. Polarizable rod-shaped molecules (biphenyls, terphenyls, etc.) are common. A common form is a pair of aromatic benzene rings, with a nonpolar moiety (pentyl, heptyl, octyl, or alkyl oxy group) on one end and polar (nitrile, halogen) on the other. Sometimes the benzene rings are separated with an acetylene group, ethylene, CH=N, CH=NO, N=N, N=NO, or ester group. In practice, eutectic mixtures of several chemicals are used, to achieve wider temperature operating range (-10..+60 °C for low-end and -20..+100 °C for high-performance displays). For example, the E7 mixture is composed of three biphenyls and one terphenyl: 39 wt.% of 4'-pentyl[1,1'-biphenyl]-4-carbonitrile (nematic range 24..35 °C), 36 wt.% of 4'-heptyl[1,1'-biphenyl]-4-carbonitrile (nematic range 30..43 °C), 16 wt.% of 4'-octoxy[1,1'-biphenyl]-4-carbonitrile (nematic range 54..80 °C), and 9 wt.% of 4-*pentyl[1,1':4',1*-terphenyl]-4-carbonitrile (nematic range 131..240 °C).

Environmental Impact

The production of LCD screens uses nitrogen trifluoride (NF_3) as an etching fluid during the production of the thin-film components. NF_3 is a potent greenhouse gas, and its relatively long half-life may make it a potentially harmful contributor to global warming. A report suggested that its effects were theoretically much greater than better-known sources of greenhouse gasses like carbon dioxide. As NF_3 was not in widespread use at the time, it was not made part of the Kyoto Protocols and has been deemed "the missing greenhouse gas".

Critics of the report point out that it assumes that all of the NF_3 produced would be released to the atmosphere. In reality, the vast majority of NF_3 is broken down during the cleaning processes; two earlier studies found that only 2 to 3% of the gas escapes destruction after its use. Furthermore, the report failed to compare NF_3's effects with what it replaced, perfluorocarbon, another powerful greenhouse gas, of which anywhere from 30 to 70% escapes to the atmosphere in typical use.

Container Glass

Container glass is a type of glass for the production of glass containers, such as bottles, jars, drinkware, and bowls. Container glass stands in contrast to *flat glass* (used for windows, glass doors, transparent walls, windshields) and *glass fiber* (used for thermal insulation, in fiberglass composites, and optical communication).

Composition

Container glass has a lower magnesium oxide and sodium oxide content than flat glass, and a higher Silica, Calcium oxide, and Aluminum oxide content. Its higher content of water-insoluble oxides imparts slightly higher chemical durability against water, which is required for storage of beverages and food.

Most container glass is soda-lime glass, produced by blowing and pressing techniques, while some laboratory glassware is made from borosilicate glass.

Glass Containers

Container glass is used in the following:

- Glass bottles:
    - Beer bottle.
    - Bologna bottle.
    - Fiasco.
    - Milk bottle.
    - Sealed bottles.
    - Wine bottles.
- Jars:
    - Antique fruit jar.
    - Killing jar.
    - Kilner jar.
    - Leyden jar.
    - Mason jar.
- Drinkware.
- Bowls.

- Pitchers.
- Vases.
- Laboratory glassware.

Mirror

A mirror is an object that reflects light in such a way that, for incident light in some range of wavelengths, the reflected light preserves many or most of the detailed physical characteristics of the original light, called specular reflection. This is different from other light-reflecting objects that do not preserve much of the original wave signal other than color and diffuse reflected light, such as flat-white paint.

The most familiar type of mirror is the plane mirror, which has a flat surface. Curved mirrors are also used, to produce magnified or diminished images or focus light or simply distort the reflected image.

Mirrors are commonly used for personal grooming or viewing oneself (where they are also called looking-glasses), for viewing the area behind and on the sides on motor vehicles while driving, for decoration, and architecture. Mirrors are also used in scientific apparatus such as telescopes and lasers, cameras, and industrial machinery. Most mirrors are designed for visible light; however, mirrors designed for other wavelengths of electromagnetic radiation are also used.

Types of Glass Mirrors

18th century vermeil mirror in the Musée des Arts décoratifs, Strasbourg.

There are many types of glass mirrors, each representing a different manufacturing process and reflection type.

An aluminium glass mirror is made of a float glass manufactured using vacuum coating, i.e. aluminium powder is evaporated (or "sputtered") onto the exposed surface of the glass in a vacuum chamber and then coated with two or more layers of waterproof protective paint.

A low aluminium glass mirror is manufactured by coating silver and two layers of protective paint on the back surface of glass. A low aluminium glass mirror is very clear, light transmissive, smooth, and reflects accurate natural colors. This type of glass is widely used for framing presentations and exhibitions in which a precise color representation of the artwork is truly essential or when the background color of the frame is predominantly white.

A safety glass mirror is made by adhering a special protective film to the back surface of a silver glass mirror, which prevents injuries in case the mirror is broken. This kind of mirror is used for furniture, doors, glass walls, commercial shelves, or public areas.

A silkscreen printed glass mirror is produced using inorganic color ink that prints patterns through a special screen onto glass. Various colors, patterns, and glass shapes are available. Such a glass mirror is durable and more moisture resistant than ordinary printed glass and can serve for over 20 years. This type of glass is widely used for decorative purposes (e.g., on mirrors, table tops, doors, windows, kitchen chop boards, etc.).

A silver glass mirror is an ordinary mirror, coated on its back surface with silver, which produces images by reflection. This kind of glass mirror is produced by coating a silver, copper film and two or more layers of waterproof paint on the back surface of float glass, which perfectly resists acid and moisture. A silver glass mirror provides clear and actual images, is quite durable, and is widely used for furniture, bathroom and other decorative purposes.

Decorative glass mirrors are usually handcrafted. A variety of shades, shapes and glass thickness are often available.

Effects

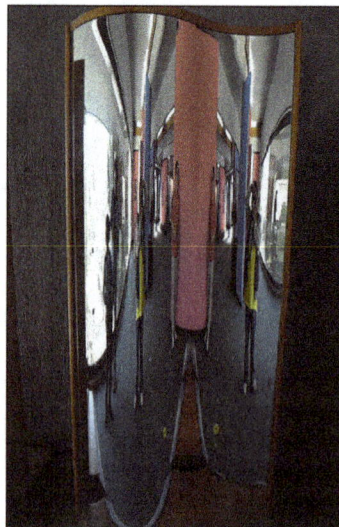

Photographer taking picture of himself in curved mirror at
the Universum museum in Mexico City.

Shape of a Mirror's Surface

A beam of light reflects off a mirror at an angle of reflection equal to its angle of incidence (if the size of a mirror is much larger than the wavelength of light). That is, if the beam of light is shining on a mirror's surface, at a $\theta°$ angle vertically, then it reflects from the point of incidence at a $\theta°$ angle, vertically in the opposite direction. This law mathematically follows from the interference of a plane wave on a flat boundary (of much larger size than the wavelength).

- In a plane mirror, a parallel beam of light changes its direction as a whole, while still remaining parallel; the images formed by a plane mirror are virtual images, of the same size as the original object.

- In a concave mirror, parallel beams of light become a convergent beam, whose rays intersect in the focus of the mirror. Also known as converging mirror.

- In a convex mirror, parallel beams become divergent, with the rays appearing to diverge from a common point of intersection "behind" the mirror.

- Spherical concave and convex mirrors do not focus parallel rays to a single point due to spherical aberration. However, the ideal of focusing to a point is a commonly used approximation. Parabolic reflectors resolve this, allowing incoming parallel rays (for example, light from a distant star) to be focused to a small spot; almost an ideal point. Parabolic reflectors are not suitable for imaging nearby objects because the light rays are not parallel.

Mirror Image

A large convex mirror. Distortions in the image
increase with the viewing distance.

Objects viewed in a (plane) mirror will appear laterally inverted (e.g., if one raises one's right hand, the image's left hand will appear to go up in the mirror), but not vertically inverted (in the image a person's head still appears above their body). However, a mirror does not usually "swap" left and right any more than it swaps top and bottom. A mirror typically reverses the forward/backward axis. To be precise, it reverses the object in the direction perpendicular to the mirror surface (the normal). Because left and right are defined relative to front-back and top-bottom, the "flipping" of front and back results in the perception of a left-right reversal in the image. (If you stand side-on to a mirror, the mirror really does reverse your left and right, because that's the direction perpendicular to the mirror).

Looking at an image of oneself with the front-back axis flipped results in the perception of an image with its left-right axis flipped. When reflected in the mirror, your right hand remains directly opposite your real right hand, but it is perceived as the left hand of your image. When a person looks into a mirror, the image is actually front-back reversed, which is an effect similar to the hollow-mask illusion. Notice that a mirror image is fundamentally different from the object and cannot be reproduced by simply rotating the object.

For things that may be considered as two-dimensional objects (like text), front-back reversal cannot usually explain the observed reversal. In the same way that text on a piece of paper appears reversed if held up to a light and viewed from behind, text held facing a mirror will appear reversed, because the observer is behind the text. Another way to understand the reversals observed in images of objects that are effectively two-dimensional is that the inversion of left and right in a mirror is due to the way human beings turn their bodies. To turn from viewing the side of the object facing the mirror to view the reflection in the mirror requires the observer to look in the opposite direction. To look in another direction, human beings turn their heads about a vertical axis. This causes a left-right reversal in the image but not an up-down reversal. If a person instead turns by bending over and looking at the mirror image between their legs, up-down will appear reversed but not left-right. This sort of reversal is simply a change relative to the observer and not a change intrinsic to the image itself, as with a three-dimensional object.

Manufacturing

Four different mirrors, showing the difference in reflectivity. Clockwise from upper left: dielectric (80%), aluminum (85%), chrome (25%), and enhanced silver (99.9%). All are first-surface mirrors except the chrome mirror. The dielectric mirror reflects yellow light from the first-surface, but acts like an antireflection coating to purple light, thus produced a ghost reflection of the lightbulb from the second-surface.

A dielectric mirror-stack works on the principle of thin-film interference. Each layer has a different refractive index, allowing each interface to produce a small amount of reflection. When the thickness of the layers is proportional to the chosen wavelength, the multiple reflections constructively interfere. Stacks may consist of a few to hundreds of individual coats.

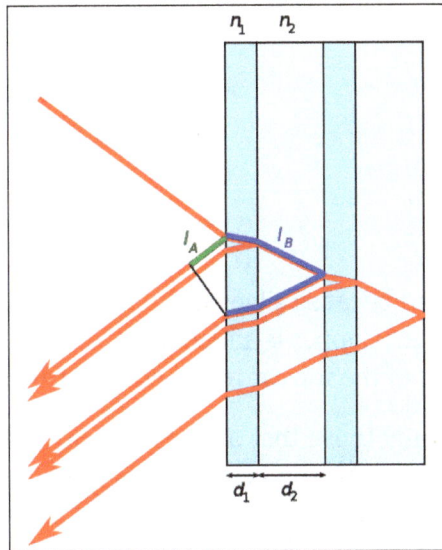

Mirrors are manufactured by applying a reflective coating to a suitable substrate. The most common substrate is glass, due to its transparency, ease of fabrication, rigidity, hardness, and ability to take a smooth finish. The reflective coating is typically applied to the back surface of the glass, so that the reflecting side of the coating is protected from corrosion and accidental damage by the glass on one side and the coating itself and optional paint for further protection on the other.

In classical antiquity, mirrors were made of solid metal (bronze, later silver) and were generally too expensive for use by common people, although during the Roman Empire even maid servants used widespread silver mirrors; they were also prone to corrosion. Due to the low reflectivity of polished metal, these mirrors also gave a darker image than modern ones, making them unsuitable for indoor use with the artificial lighting of the time (candles or lanterns).

The method of making mirrors out of plate glass was invented by 13th-century Venetian glassmakers on the island of Murano, who covered the back of the glass with an amorphous coat of tin using a fire-gilding technique, obtaining near-perfect and undistorted reflection. For over one hundred years, Venetian mirrors installed in richly decorated frames served as luxury decorations for palaces throughout Europe, but the secret of the mercury process eventually arrived in London and Paris during the

17th century, due to industrial espionage. French workshops succeeded in large-scale industrialization of the process, eventually making mirrors affordable to the masses, although mercury's toxicity (a primary ingredient in gilding, which was boiled away forming noxious vapors) remained a problem.

In modern times, the mirror substrate is shaped, polished and cleaned, and is then coated. Glass mirrors are most often coated with silver or aluminium, implemented by a series of coatings:

- Tin(II) chloride.
- Silver.
- Chemical activator.
- Copper.
- Paint.

The tin(II) chloride is applied because silver will not bond with the glass. The activator causes the tin/silver to harden. Copper is added for long-term durability. The paint protects the coating on the back of the mirror from scratches and other accidental damage.

In some applications, generally those that are cost-sensitive or that require great durability, such as for mounting in a prison cell, mirrors may be made from a single, bulk material such as polished metal. However, metals consist of small crystals (grains) separated by grain boundaries. Thus, crystalline metals do not reflect with perfect uniformity. Other methods like wet-deposition or electroplating produce a non-crystalline coating of amorphous metal (metallic glass). Lacking any grain boundaries, the amorphous coatings have higher reflectivity than crystalline metals of the same type. Electroplating must be performed by first coating the glass with carbon, to make the surface electrically conductive, thus the adhesion is often not as good as with wet-deposition. Both lack the ability to produce perfectly uniform thicknesses with high precision. When high precision or reflectivity is not a requirement, the coating may be placed on the back of the mirror so that the light passes through the glass, and the coating is the second surface it encounters. Therefore, these are called second-surface mirrors, which have the added benefit of high durability, because the glass substrate can protect the coating from damage.

For technical applications such as laser mirrors, the reflective coating is typically applied by vacuum deposition. Vacuum deposition provides an effective means of producing a very uniform coating, and controlling the thickness with high precision. In applications where great precision and low losses are required, the coated side of the mirror may be the first material encountered by the light, referred to as a first-surface mirror. This eliminates refraction and double reflections, also called "ghost reflections" (a weak reflection from the surface of the glass, and a stronger one from the reflecting metal), and reduces absorption of light by the mirror. Technical mirrors may use a silver, aluminium, or gold coating (the latter typically for infrared mirrors), and achieve reflectivities of 90–95% when new. A hard, protective, transparent overcoat may be applied to prevent oxidation of the reflective layer and scratching of the soft metal.

Applications requiring higher reflectivity or greater durability, where wide bandwidth is not essential, use dielectric coatings, which can achieve reflectivities as high as 99.997% over a limited range of wavelengths. Because they are often chemically stable and do not conduct electricity, dielectric coatings are almost always applied by methods of vacuum deposition, and most commonly by evaporation deposition. Because the coatings are usually transparent, absorption losses are negligible. Unlike with metals, the reflectivity of the individual dielectric-coatings is a function of Snell's law known as the Fresnel equations, determined by the difference in refractive index between layers. Therefore, the thickness and index of the coatings can be adjusted to be centered on any wavelength. Vacuum deposition can be achieved in a number of ways, including sputtering, evaporation deposition, arc deposition, reactive-gas deposition, and ion plating, among many others.

Tolerances

Polishing the primary mirror for the Hubble Space Telescope. A deviation in the surface quality of approximately 4λ resulted in poor images initially, which was eventually compensated for using corrective optics.

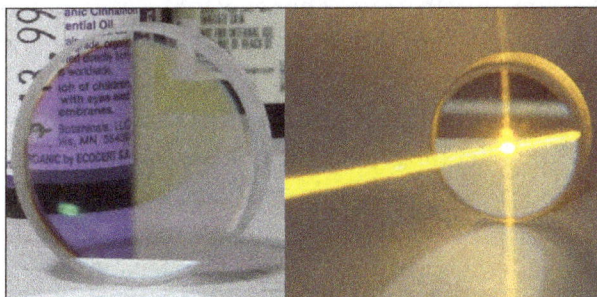

A dielectric, laser output-coupler that is 75–80% reflective between 500 and 600 nm. Left: The mirror is highly reflective to yellow and green but highly transmissive to red and blue. Right: The mirror transmits 25% of the 589 nm laser light. Because the smoke particles diffract more light than they reflect, the beam appears much brighter when reflecting back toward the observer.

Flatness errors, like rippled dunes across the surface, produced these artifacts, distortion, and low image quality in the far field reflection of a household mirror.

Mirrors can be manufactured to a wide range of engineering tolerances, including reflectivity, surface quality, surface roughness, or transmissivity, depending on the desired application. These tolerances can range from low, such as found in a normal household-mirror, to extremely high, like those used in lasers or telescopes. Increasing the tolerances allows better and more precise imaging or beam transmission over longer distances. In imaging systems this can help reduce anomalies (artifacts), distortion or blur, but at a much higher cost. Where viewing distances are relatively close or high precision is not a concern, lower tolerances can be used to make effective mirrors at affordable costs.

Reflectivity

The reflectivity of a mirror is determined by the percentage of reflected light per the total of the incident light. The reflectivity may vary with wavelength. All or a portion of the light not reflected is absorbed by the mirror, while in some cases a portion may also transmit through. Although some small portion of the light will be absorbed by the coating, the reflectivity is usually higher for first-surface mirrors, eliminating both reflection and absorption losses from the substrate. The reflectivity is often determined by the type and thickness of the coating. When the thickness of the coating is sufficient to prevent transmission, all of the losses occur due to absorption. Aluminum is harder, less expensive, and more resistant to tarnishing than silver, and will reflect 85 to 90% of the light in the visible to near-ultraviolet range, but experiences a drop in its reflectance between 800 and 900 nm. Gold is very soft and easily scratched, costly, yet does not tarnish. Gold is greater than 96% reflective to near and far-infrared light between 800 and 12000 nm, but poorly reflects visible light with wavelengths shorter than 600 nm (yellow). Silver is expensive, soft, and quickly tarnishes, but has the highest reflectivity in the visual to near-infrared of any metal. Silver can reflect up to 98 or 99% of light to wavelengths as long as 2000 nm, but loses nearly all reflectivity at wavelengths shorter than 350 nm.

Dielectric mirrors can reflect greater than 99.99% of light, but only for a narrow range of wavelengths, ranging from a bandwidth of only 10 nm to as wide as 100 nm for tunable lasers. However, dielectric coatings can also enhance the reflectivity of metallic coatings and protect them from scratching or tarnishing. Dielectric materials are typically very hard and relatively cheap, however the number of coats needed generally makes it an expensive process. In mirrors with low tolerances, the coating thickness may be reduced to save cost, and simply covered with paint to absorb transmission.

Surface Quality

Surface quality, or surface accuracy, measures the deviations from a perfect, ideal surface shape. Increasing the surface quality reduces distortion, artifacts, and aberration in images, and helps increase coherence, collimation, and reduce unwanted divergence in beams. For plane mirrors, this is often described in terms of flatness, while other surface shapes are compared to an ideal shape. The surface quality is typically measured with items like interferometers or optical flats, and are usually measured in wavelengths of light (λ). These deviations can be much larger or much smaller than the surface roughness. A normal household-mirror made with float glass may have flatness tolerances as low as 9–14λ per inch (25.4 mm), equating to a deviation of 5600 through 8800 nanometers from perfect flatness. Precision ground and polished mirrors intended for lasers or telescopes may have tolerances as high as $\lambda/50$ (1/50 of the wavelength of the light, or around 12 nm) across the entire surface. The surface quality can be affected by factors such as temperature changes, internal stress in the substrate, or even bending effects that occur when combining materials with different coefficients of thermal expansion, similar to a bimetallic strip.

Surface Roughness

Surface roughness describes the texture of the surface, often in terms of the depth of the microscopic scratches left by the polishing operations. Surface roughness determines how much of the reflection is specular and how much diffuses, controlling how sharp or blurry the image will be.

For perfectly specular reflection, the surface roughness must be kept smaller than the wavelength of the light. Microwaves, which sometimes have a wavelength greater than an inch (~25 mm) can reflect specularly off a metal screen-door, continental ice-sheets, or desert sand, while visible light, having wavelengths of only a few hundred nanometers (a few hundred-thousandths of an inch), must meet a very smooth surface to produce specular reflection. For wavelengths that are approaching or are even shorter than the diameter of the atoms, such as X-rays, specular reflection can only be produced by surfaces that are at a grazing incidence from the rays.

Surface roughness is typically measured in microns, wavelength, or grit size, with ~80,000–100,000 grit or ~½λ–¼λ being "optical quality".

Transmissivity

Transmissivity is determined by the percentage of light transmitted per the incident light. Transmissivity is usually the same from both first and second surfaces. The combined transmitted and reflected light, subtracted from the incident light, measures the amount absorbed by both the coating and substrate. For transmissive mirrors, such as one-way mirrors, beam splitters, or laser output couplers, the transmissivity of the mirror is an important consideration. The transmissivity of metallic coatings are often determined by their thickness. For precision beam-splitters or output couplers, the thickness of the coating must be kept at very high tolerances to transmit the proper amount of light. For dielectric mirrors, the thickness of the coat must always be kept to high tolerances, but it is often more the number of individual coats that determine the transmissivity. For the substrate, the material used must also have good transmissivity to the chosen wavelengths. Glass is a suitable substrate for most visible-light applications, but other substrates such as zinc selenide or synthetic sapphire may be used for infrared or ultraviolet wavelengths.

Wedge

Wedge errors are caused by the deviation of the surfaces from perfect parallelism. An optical wedge is the angle formed between two plane-surfaces (or between the principle planes of curved surfaces) due to manufacturing errors or limitations, causing one edge of the mirror to be slightly thicker than the other. Nearly all mirrors and optics with parallel faces have some slight degree of wedge, which is usually measured in seconds or minutes of arc. For first-surface mirrors, wedges can introduce alignment deviations in mounting hardware. For second-surface or transmissive mirrors, wedges can have a prismatic effect on the light, deviating its trajectory or, to a very slight degree, its color, causing chromatic and other forms of aberration. In some instances, a slight wedge is desirable, such as in certain laser systems where stray reflections from the uncoated surface are better dispersed than reflected back through the medium.

Surface Defects

Surface defects are small-scale, discontinuous imperfections in the surface smoothness. Surface defects are larger (in some cases much larger) than the surface roughness, but only affect small, localized portions of the entire surface. These are typically found as scratches, digs, pits (often from bubbles in the glass), sleeks (scratches from prior, larger grit polishing operations that were not fully removed by subsequent polishing grits), edge chips, or blemishes in the coating. These defects are often an unavoidable side-effect of manufacturing limitations, both in cost and machine precision. If kept low enough, in most applications these defects will rarely have any adverse effect, unless the surface is located at an image plane where they will show up directly. For applications that require extremely low scattering of light, extremely high reflectance, or low absorption due to high energy-levels that could destroy the mirror, such as lasers or Fabry-Perot interferometers, the surface defects must be kept to a minimum.

Applications

A cheval glass.

Reflections in a spherical convex mirror.
The photographer is seen at top right.

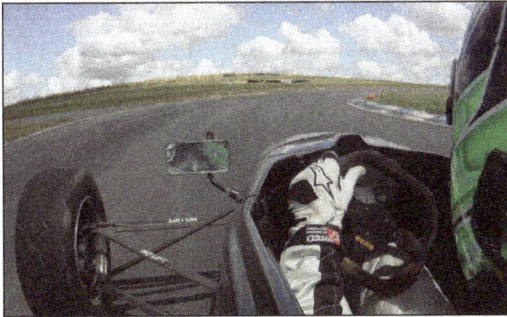

A side-mirror on a racing car.

Rear view mirror.

Personal Grooming

Mirrors are commonly used as aids to personal grooming. They may range from small sizes, good to carry with oneself, to full body sized; they may be handheld, mobile, fixed or adjustable. A classic example of the latter is the cheval glass, which may be tilted.

Safety and Easier Viewing

Convex mirrors: Convex mirrors provide a wider field of view than flat mirrors, and are often used on vehicles, especially large trucks, to minimize blind spots. They are sometimes placed at road junctions, and corners of sites such as parking lots to allow people to see around corners to avoid crashing into other vehicles or shopping carts. They are also sometimes used as part of security systems, so that a single video camera can show more than one angle at a time. Convex mirrors as decoration are used in interior design to provide a predominantly experiential effect.

Mouth mirrors or "dental mirrors": Mouth mirrors or "dental mirrors" are used by dentists to allow indirect vision and lighting within the mouth. Their reflective surfaces may be either flat or curved. Mouth mirrors are also commonly used by mechanics to allow vision in tight spaces and around corners in equipment.

Rear-view mirrors: Rear-view mirrors are widely used in and on vehicles (such as automobiles, or bicycles), to allow drivers to see other vehicles coming up behind them. On rear-view sunglasses, the left end of the left glass and the right end of the right glass work as mirrors.

One-way Mirrors and Windows

One-way mirrors: One-way mirrors (also called two-way mirrors) work by overwhelming dim transmitted light with bright reflected light. A true one-way mirror that actually allows light to be transmitted in one direction only without requiring external energy is not possible as it violates the second law of thermodynamics.

One-way windows: One-way windows can be made to work with polarized light in the laboratory without violating the second law. This is an apparent paradox that stumped some great physicists, although it does not allow a practical one-way mirror for use in the real world. Optical isolators are one-way devices that are commonly used with lasers.

Signalling

With the sun as light source, a mirror can be used to signal by variations in the orientation of the mirror. The signal can be used over long distances, possibly up to 60 kilometres (37 mi) on a clear day. This technique was used by Native American tribes and numerous militaries to transmit information between distant outposts.

Mirrors can also be used for search to attract the attention of search and rescue parties. Specialized type of mirrors are available and are often included in military survival kits.

Technology

Televisions and Projectors

Microscopic mirrors are a core element of many of the largest high-definition televisions and video projectors. A common technology of this type is Texas Instruments' DLP. A DLP chip is a postage stamp-sized microchip whose surface is an array of millions of microscopic mirrors. The picture is created as the individual mirrors move to either reflect light toward the projection surface (pixel on), or toward a light absorbing surface (pixel off).

Other projection technologies involving mirrors include LCoS. Like a DLP chip, LCoS is a microchip of similar size, but rather than millions of individual mirrors, there is a single mirror that is actively shielded by a liquid crystal matrix with up to millions of pixels. The picture, formed as light, is either reflected toward the projection surface (pixel on), or absorbed by the activated LCD pixels (pixel off). LCoS-based televisions and projectors often use 3 chips, one for each primary color.

Large mirrors are used in rear projection televisions. Light is "folded" by one or more mirrors so that the television set is compact.

Solar Power

Parabolic troughs near Harper Lake in California.

Mirrors are integral parts of a solar power plant. The one shown in the adjacent picture uses concentrated solar power from an array of parabolic troughs.

Instruments

E-ELT mirror segments under test.

Telescopes and other precision instruments use *front silvered* or first surface mirrors, where the reflecting surface is placed on the front (or first) surface of the glass (this eliminates reflection from glass surface ordinary back mirrors have). Some of them use silver, but most are aluminium, which is more reflective at short wavelengths than silver. All of these coatings are easily damaged and require special handling. They reflect 90% to 95% of the incident light when new. The coatings are typically applied by vacuum deposition. A protective overcoat is usually applied before the mirror is removed from the vacuum, because the coating otherwise begins to corrode as soon as it is ex-posed to oxygen and humidity in the air. *Front silvered* mirrors have to be resurfaced occasionally to keep their quality. There are optical mirrors such as mangin mirrors that are *second surface mirrors* (reflective coating on the rear surface) as part of their optical designs, usually to correct optical aberrations.

Deformable thin-shell mirror. It is 1120 millimetres across but just 2 millimetres thick, making it much thinner than most glass windows.

The reflectivity of the mirror coating can be measured using a reflectometer and for a particular metal it will be different for different wavelengths of light. This is exploited in some optical work to make cold mirrors and hot mirrors. A cold mirror is made by using a transparent substrate and choosing a coating material that is more reflective to visible light and more transmissive to infrared light.

A dielectric coated mirror used in a dye laser. The mirror is over 99% reflective at 550 nanometers, (yellow), but will allow most other colors to pass through.

A hot mirror is the opposite, the coating preferentially reflects infrared. Mirror surfaces are sometimes given thin film overcoatings both to retard degradation of the surface and to increase their reflectivity in parts of the spectrum where they will be used. For instance, aluminum mirrors are commonly coated with silicon dioxide or magnesium fluoride. The reflectivity as a function of wavelength depends on both the thickness of the coating and on how it is applied.

A dielectric mirror used in tunable lasers. With a center wavelength of 600 nm and bandwidth of 100 nm, the coating is totally reflective to the orange construction paper, but only reflects the reddish hues from the blue paper.

For scientific optical work, dielectric mirrors are often used. These are glass (or sometimes other material) substrates on which one or more layers of dielectric material are deposited, to form an optical coating. By careful choice of the type and thickness of the dielectric layers, the range of wavelengths and amount of light reflected from the mirror can be specified. The best mirrors of this type can reflect >99.999% of the light (in a narrow range of wavelengths) which is incident on the mirror. Such mirrors are often used in lasers.

In astronomy, adaptive optics is a technique to measure variable image distortions and adapt a deformable mirror accordingly on a timescale of milliseconds, to compensate for the distortions.

Although most mirrors are designed to reflect visible light, surfaces reflecting other forms of electromagnetic radiation are also called "mirrors". The mirrors for other ranges of electromagnetic waves are used in optics and astronomy. Mirrors for radio waves (sometimes known as reflectors) are important elements of radio telescopes.

Face-to-face Mirrors

Two or more mirrors aligned exactly parallel and facing each other can give an infinite regress of reflections, called an infinity mirror effect. Some devices use this to generate multiple reflections:

- Fabry–Pérot interferometer.
- Laser (which contains an optical cavity).
- 3D Kaleidoscope to concentrate light.
- Momentum-enhanced solar sail.

Military Applications

It has been said that Archimedes used a large array of mirrors to burn Roman ships during an attack on Syracuse. This has never been proven or disproved; however, it has been put to the test. Recently, on a popular Discovery Channel show, *Myth-Busters*, a team from MIT tried to recreate the famous "Archimedes Death Ray". They were unsuccessful at starting a fire on the ship. Previous attempts to light the boat on fire using only the bronze mirrors available in Archimedes' time were unsuccessful, and the time taken to ignite the craft would have made its use impractical, resulting in the MythBusters team deeming the myth "busted". It was however found that the mirrors made it very difficult for the passengers of the targeted boat to see, likely helping to cause their defeat, which may have been the origin of the myth.

Seasonal Lighting

A multi-facet mirror in the Kibble Palace conservatory, Glasgow, Scotland.

Due to its location in a steep-sided valley, the Italian town of Viganella gets no direct sunlight for seven weeks each winter. In 2006 a €100,000 computer-controlled mirror, 8×5 m, was installed to reflect sunlight into the town's piazza. In early 2007 the similarly situated village of Bondo, Switzerland, was considering applying this solution as well. In 2013, mirrors were installed to reflect sunlight into the town square in the Norwegian town of Rjukan. Mirrors can be used to produce enhanced lighting effects in greenhouses or conservatories.

Architecture

Mirrors are a popular design theme in architecture, particularly with late modern and post-modernist high-rise buildings in major cities. Early examples include the Campbell Center in Dallas, which opened in 1972, and the John Hancock Tower in Boston.

Mirrored building in Manhattan.

More recently, two skyscrapers designed by architect Rafael Viñoly, the Vdara in Las Vegas and 20 Fenchurch Street in London, have experienced unusual problems due to their concave curved glass exteriors acting as respectively cylindrical and spherical reflectors for sunlight. In 2010, a study reported that sunlight reflected off the Vdara's south-facing tower could singe swimmers in the hotel pool, as well as melting plastic cups and shopping bags; employees of the hotel referred to the phenomenon as the "Vdara death ray", aka the "fryscraper." In 2013, sunlight reflecting off 20 Fenchurch Street melted parts of a Jaguar car parked nearby and scorching or igniting the carpet of a nearby barber shop. This building had been nicknamed the "walkie-talkie" because its shape was supposedly similar to a certain model of two-way radio; but after its tendency to overheat surrounding objects became known, the nickname changed to the "walkie-scorchie."

Fine art

Paintings

Painters depicting someone gazing into a mirror often also show the person's reflection. This is a kind of abstraction—in most cases the angle of view is such that the person's reflection should not be visible. Similarly, in movies and still photography an actor or actress is often shown ostensibly looking at him- or herself in the mirror, and yet the reflection faces the camera. In reality, the actor or actress sees only the camera and its operator in this case, not their own reflection.

The mirror is the central device in some of the greatest of European paintings:

- Édouard Manet's *A Bar at the Folies-Bergère*.

- Titian's *Venus with a Mirror*.

- Jan van Eyck's *Arnolfini Portrait*.

- Pablo Picasso's *Girl before a Mirror*.

- Diego Velázquez's *Las Meninas*, wherein the viewer is both the watcher (of a self-portrait in progress) and the watched, and the many adaptations of that painting in various media.

- Veronese's *Venus with a Mirror*.

Titian's *Venus with a Mirror*.

Mirrors have been used by artists to create works and hone their craft:

- Filippo Brunelleschi discovered linear perspective with the help of the mirror.

- Leonardo da Vinci called the mirror the "master of painters". He recommended, "When you wish to see whether your whole picture accords with what you have portrayed from nature take a mirror and reflect the actual object in it. Compare what is reflected with your painting and carefully consider whether both likenesses of the subject correspond, particularly in regard to the mirror."

- Many self-portraits are made possible through the use of mirrors, such as the great self-portraits by Dürer, Frida Kahlo, Rembrandt, and Van Gogh. M. C. Escher used special shapes of mirrors in order to achieve a much more complete view of his surroundings than by direct observation in *Hand with Reflecting Sphere* (also known as *Self-Portrait in Spherical Mirror*).

Mirrors are sometimes necessary to fully appreciate art work:

- István Orosz's anamorphic works are images distorted such that they only become clearly visible when reflected in a suitably shaped and positioned mirror.

Sculpture

Mirrors in interior design: Waiting room in the house of M.me B.

- Anamorphosis projecting sculpture into mirrors:

  Contemporary anamorphic artist Jonty Hurwitz uses cylindrical mirrors to project distorted sculptures.

- Sculptures comprised entirely or in part of mirrors:

  ◦ *Infinity Also Hurts* is a mirror, glass and silicone sculpture by artist, Seth Wulsin.

  ◦ *Sky Mirror* is a public sculpture by artist, Anish Kapoor.

Other Artistic Mediums

Grove of Mirrors.

Some other contemporary artists use mirrors as the material of art:

- A Chinese magic mirror is an art in which the face of the bronze mirror projects the same image that was cast on its back. This is due to minute curvatures on its front.

- Specular holography uses a large number of curved mirrors embedded in a surface to produce three-dimensional imagery.

- Paintings on mirror surfaces (such as silkscreen printed glass mirrors).

- Special mirror installations:

    o *Follow Me* mirror labyrinth by artist, Jeppe Hein.

    o *Mirror Neon Cube* by artist, Jeppe Hein.

Decoration

Chimneypiece and overmantel mirror.

Glasses with mirrors – Prezi HQ.

Mirrors are frequently used in interior decoration and as ornaments:

- Mirrors, typically large and unframed, are frequently used in interior decoration to create an illusion of space and amplify the apparent size of a room. They come also framed in a variety of forms, such as the pier glass and the overmantel mirror.

- Mirrors are used also in some schools of feng shui, an ancient Chinese practice of placement and arrangement of space to achieve harmony with the environment.

- The softness of old mirrors is sometimes replicated by contemporary artisans for use in interior design. These reproduction antiqued mirrors are works of art and can bring color and texture to an otherwise hard, cold reflective surface. It is an artistic process that has been attempted by many and perfected by few.

- A decorative reflecting sphere of thin metal-coated glass, working as a reducing wide-angle mirror, is sold as a Christmas ornament called a *bauble*.

Mirrors and Animals

Only a few animal species have been shown to have the ability to recognize themselves in a mirror, most of them mammals. Experiments have found that the following animals can pass the mirror test:

- All great apes:
  - Humans. Humans tend to fail the mirror test until they are about 18 months old, or what psychoanalysts call the "mirror stage".
  - Bonobos.
  - Chimpanzees.
  - Orangutans.
  - Gorillas. Initially, it was thought that gorillas did not pass the test, but there are now several well-documented reports of gorillas (such as Koko) passing the test.
- Bottlenose dolphins.
- Orcas.
- Elephants.
- European magpies.

Other Types of Mirrors

4.5-metre (15 ft) high acoustic mirror near Kilnsea Grange, East Yorkshire, UK.

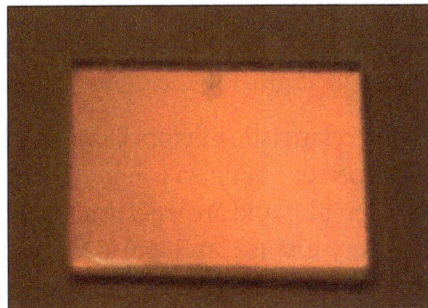

A hot mirror used in a camera to reduce red eye.

Other types of reflecting device are also called mirrors.

- Acoustic mirrors are passive devices used to reflect and perhaps to focus sound waves. Acoustic mirrors were used for selective detection of sound waves, especially during World War II. They were used for detection of enemy aircraft, prior to the development of radar. Acoustic mirrors are used for remote probing of the atmosphere; they can be used to form a narrow diffraction-limited beam. They can also be used for underwater imaging.

- Active mirrors are mirrors that amplify the light they reflect. They are used to make disk lasers. The amplification is typically over a narrow range of wavelengths, and requires an external source of power.

- Atomic mirrors are devices which reflect matter waves. Usually, atomic mirrors work at grazing incidence. Such mirrors can be used for atomic interferometry and atomic holography. It has been proposed that they can be used for non-destructive imaging systems with nanometer resolution.

- Cold mirrors are dielectric mirrors that reflect the entire visible light spectrum, while efficiently transmitting infrared wavelengths. These are the converse of hot mirrors.

- Corner reflectors use three flat mirrors to reflect light back towards its source, they may also be implemented with prisms that reflect using total internal reflection that have no mirror surfaces. They are used for emergency location, and traffic reflectors.

- Hot mirrors reflect infrared light while allowing visible light to pass. These can be used to separate useful light from unneeded infrared to reduce heating of components in an optical device. They can also be used as dichroic beamsplitters. (Hot mirrors are the converse of cold mirrors).

- Metallic reflectors are used to reflect infrared light (such as in space heaters or microwaves).

- Non-reversing mirrors are mirrors that provide a non-reversed image of their subjects.

- X-ray mirrors produce specular reflection of X-rays. All known types work only at angles near grazing incidence, and only a small fraction of the rays are reflected.

- Flying relativistic mirrors use a shockwave (wake wave) in a plasma to produce not only specular but also coherent reflection of high-energy radiation with very short wavelengths, and are used in some types of X-ray lasers. The wake wave is produced by a very intense laser-pulse focused into the low-density plasma, which creates a wall of charged particles and free electrons moving at an extremely high velocity. Based on the Theory of Special Relativity, because of the wake wave's speed, it can reflect radiation with wavelengths much shorter than visible or UV light.

Glass-ceramic

Glass ceramic materials have the same chemical compositions as glasses but differ from them in that they are typically 95-98% crystalline by volume, with only a small percentage vitreous. The crystals themselves are generally very small, less than 1μm and most often very uniform in size. Furthermore, due to their crystallinity and network of grain boundaries, they are no longer transparent.

Properties of Glass-ceramic Materials

Glass ceramic materials are typically characterised by:

- High strength.

- High impact resistance.

- Low co-efficient of thermal expansion, sometimes even negative co-efficient of thermal expansion.

- Good resistance to thermal shock.

- A range of optical properties, from translucent to opaque and sometime opalescence.

Production of Glass-ceramic Materials

Glass ceramic components are formed using the same processes that are applicable to glass components. To convert them from a vitreous glass material into a crystalline glass ceramic material they must be heat treated or devitrified.

Devitrification can occur spontaneously during cooling or in service, but is most commonly incorporated to produce glass ceramics. It involves heating the formed glass product to a temperature high enough to stimulate crystals to nucleate throughout the glass. The temperature is then increased, which induces growth of the nuclei, crystallising the remaining glass.

Nucleation requires a critical number of atoms converging to form a nucleus. When the nucleus reaches critical size, nucleation occurs. In many glass compositions, nucleation is hampered by the fact the material is silica-based and highly viscous, making it difficult for the required atoms to come together. The crystal compositions can also be complex making nucleation difficult. These factors aid glass forming and cooling without crystallisation.

The devitrification heat treatment must be carefully controlled to ensure the maximum number of nuclei are formed and that these nuclei grow into a uniform fine crystal structure. In order to obtain a high concentration of nuclei throughout the structure, it is common to add a nucleating agent to the glass composition.

Nucleating Agents

The most common nucleating agents are TiO_2 and ZrO_2. Other materials that have been used for nucleating agents include P_2O_5, platinum group and noble metals and some fluorides.

Glass-ceramic Compositions

While many different glass ceramic compositions exist, there are 3 main families:

- LAS: A mixture of lithium, aluminium and silicon oxides (Li_2O-Al_2O_3-SiO_2), with other glass forming agents (e.g. Na_2O, K_2O and CaO).

- MAS: A mixture of magnesium, aluminium and silicon oxides (MgO- Al_2O_3- SiO_2) with glass forming agents.

- ZAS: A mixture of zinc, aluminium and silicon oxides (ZnO- Al_2O_3-SiO_2) with glass forming agents.

Applications of Glass-ceramics

Some applications of glass ceramics include:

- Radomes: Made from Corning 9606 ($2MgO.2\ Al_2O_3$, cordierite system).

- Cookware, bakeware and cooktops: Made from Corning 9608 (ß-spodumene system).

- Telescopic mirrors: Made from Owens-Illinois Cer-Vit (ß-quartz system).

- Insulators: Made from General Electric Re-X ($Li_2O.2SiO_2$ system).

- Bioactive glass for biomaterials: Bioglass 45S5 (46.1 mol% SiO_2, 26.9 mol% CaO, 24.4 mol% Na_2O and 2.5 mol% P_2O_5) has been FDA approved and many different variations on this composition have resulted.

- Engineering components: Made from Macor, a machineable glass ceramic.

Art Glass

Art glass is an item that is made, generally as an artwork for decoration but often also for utility, from glass, sometimes combined with other materials. Techniques include stained glass windows, leaded lights (also called leadlights), glass that has been placed into a kiln so that it will mould into a shape, glassblowing, sandblasted glass, and copper-foil glasswork. In general the term is restricted to relatively modern pieces made by people who see themselves as artists who have chosen to work in the medium of glass and both design and make their own pieces as fine art, rather than traditional glassworker craftsmen, who often produce pieces designed by others, though their pieces certainly may form part of art. Studio glass is another term often used for modern glass made for artistic purposes. Art glass has grown in popularity in recent years with many artists becoming famous for their work; and, as a result, more colleges are offering courses in glass work.

Techniques and Processes

Stained Glass

Stained-glass windows at Sainte Chapelle, Paris.

Stained glass, such as the windows that are seen in churches, are windows that contain an element of painting in them. The window is designed. After the glass has been cut to shape, paint that contains ground glass is applied, so that, when it is fired in a kiln, the paint fuses onto the glass surface. Following this process, the sections of glass are placed together and held in place with lead came that is then soldered at the joints. Leadlights and stained glass are manufactured in the same way, but leadlights do not contain any sections of glass that have been painted.

Blown Glass

Glassblowing is one of the most used technique for creating "art glass" and is still favoured by most of today's studio glass artists. This is because of the artist's intimacy with the material, and an almost infinite opportunity for creativity and variation at almost every stage of the process. Glassblowing can be used to create a multitude of shapes and can incorporate color through a wide range of techniques. Coloured glass can be gathered out of a crucible, clear glass can be rolled in powdered colored glass to coat the outside of a bubble, it can be rolled in chips of glass, it can be stretched into rods and incorporated through caneworking, or it can be layered, cut and fused into tiles, and incorporated into a bubble of glass for intricate patterns through murrine. "Blown glass" refers only to individually hand-made items but can include the use of moulds for shaping, ribbing, and spiking to produce decorative bubbles. Glass blown articles must be made of compatible glass or the stress in the piece will cause a failure.

Kiln-formed Glass

Kiln formed glass is usually referred to as warm glass, and can be either made up from a single piece of glass that is slumped into or over a mould or different colours and sheets of glass fused together. The process of hot glass is highly scientific in that the types of glass and temperatures that they must be fired at is quite complicated operation to undertake correctly. Art glass that is kiln formed usually take the form of dishes, plates or tiles. Glass that is fused in a kiln must be of the same co-efficient of expansion (CoE). If glass that does not have the same CoE is used for fusing, the differing rates of contraction will cause minute stress fractures to form and, over time, these fractures will cause a piece to crack. The use of polarizing filters to inspect the work will determine if stress fractures are present.

Cold Glass

Cold glass is worked by any method that does not use heat. Processes include sandblasting, cutting, sawing, chiseling, bonding and gluing.

Sandblasting

Glass can be decorated by sandblasting the surface of a piece in order to remove a layer of glass, thereby making a design stand out. Items that are sandblasted are usually thick slabs of glass into which a design has been carved by means of high pressure sandblasting. This technique provides a three-dimensional effect but is not suitable for toughened glass as the process could shatter it.

Copperfoil Technique

Louis Comfort Tiffany vase.

The technique of using copperfoil is mainly used in the construction of smaller pieces such as Tiffany style lamps, and it was, in fact, frequently used by Louis Comfort Tiffany. It consists of wrapping cut sections of glass in a self-adhesive tape that is made out of thin copper foil. This technique requires a great deal of dexterity and is also very time-consuming. After the sections have been foiled, they are soldered together in order to form the item.

Factory Art Glass

Most antique art glass was made in factories, particularly in the UK, the United States, and Bohemia, where items were made to a standard, or "pattern". This would seem contrary to the idea that art glass is distinctive and shows individual skill. However, the importance of decoration – in the Victorian era in particular – meant that much of the artistry lay with the decorator. Any assumption today that factory-made items were necessarily made by machine is incorrect. Up to about 1940, most of the processes involved in making decorative art glass were performed by hand.

Factory Differentiation and Distinctiveness

Manufacturers got around the problem of an inherent similarity in their products in various ways. First, they would frequently change designs according to demand. This was especially so in the export-dependent factories of Bohemia where salesmen would report sales trends back to the factory during each trip. Second, the decoration for mid- and lower-market items, often done by contracted "piece" workers, was often a variation on a theme. Such was the skill of these subcontractors that a reasonable standard of quality and a high rate of output were generally maintained. Finally, a high degree of differentiation could be gained from the multiplication of shapes, colours, and decorative designs, yielding many different combinations. Concurrently, from the same factories came distinctive, artistic items produced in more limited quantities for the upper-market consumer. These were decorated in-house where decorators could work closely with designers and management in order to produce a piece that was profitable.

Usable Art Glass

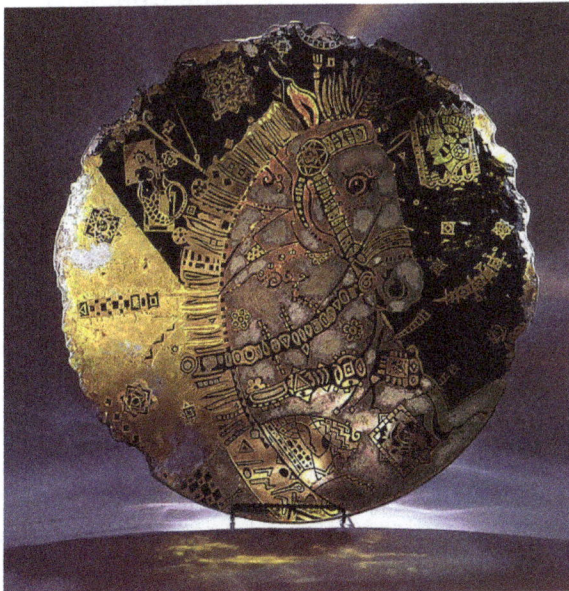

"Imperial Glass Bowl" Moulded glass art by Csaba Markus.

Many items that are now considered art glass were originally intended for use. Often that use has ceased to be relevant, but even if not, in the Victorian era and for some decades beyond useful items were often decorated to such a high degree that we can now appreciate them for their artistic or design merits.

Some art glass retains its original purpose but has come to be appreciated more for its art than for its use. Collectors of antique perfume bottles, for example, tend to display their items empty. As items of packaging, these bottles would originally have been used and thus would not ordinarily have been considered art glass. However, because of fashion trends, then as now, producers supplied goods in beautiful packaging. Lalique's Art Nouveau and Art Deco designs and Joseph Hoffman's Art Deco designs have come to be considered art glass due to their stylish and highly original decorative designs.

Decorating Techniques

- Colour: Various colours inter-mixed or otherwise incorporated.

- Texture: Frosting, satinizing, glue-chip, overshot and sandblasting.

- Surfaces: Overlays, cameo, cut-back, cutting and engraving.

Refined Glassware

Up-market refined glassware, usually lead crystal, is highly decorated and is revered for its high quality of workmanship, the purity of the metal (molten glass mixture), and the decorative techniques used, most often cutting and gilding. Both techniques continue to be used in the decoration of many pieces made from lead crystal, and nowadays these pieces are regarded as art glass.

Cut Glass

Cut glass is most often produced by hand, but automation is now becoming more common. Some designs show artistic flair, but most tend to be regular, geometric, and repetitious. Occasionally, the design can be considered a "pattern" to be replicated as exactly as possible, with the main purpose being to accentuate the refractive qualities, or "sparkle", of the crystal – certainly an aesthetic consideration, but not generally considered artistic.

Art Cut

A clear exception could be made for the highly distinctive cut crystal designs which were produced in limited quantities by designers of note. Examples are the designs of Keith Murray for Steven & Williams and those of Clyne Farquharson for John Walsh Walsh. A relatively new term is coming into use for this genre: "Art Cut".

Windshield

The windshield (North American English) or windscreen (Commonwealth English) of an aircraft, car, bus, motorbike or tram is the front window, which provides visibility while protecting occupants from the elements. Modern windshields are generally made of laminated safety glass, a type of treated glass, which consists of, typically, two curved sheets of glass with a plastic layer laminated between them for safety, and bonded into the window frame.

Motorbike windshields are often made of high-impact polycarbonate or acrylic plastic.

Usage

Windshields protect the vehicle's occupants from wind and flying debris such as dust, insects, and rocks, and provide an aerodynamically formed window towards the front. UV coating may be applied to screen out harmful ultraviolet radiation. However, this is usually unnecessary since most auto windshields are made from laminated safety glass. The majority of UV-B is absorbed by the glass itself, and any remaining UV-B together with most of the UV-A is absorbed by the PVB bonding layer.

Split and raked windshield on a 1952 DeSoto. Note the panes of glass are flat.

On motorbikes their main function is to shield the rider from wind, though not as completely as in a car, whereas on sports and racing motorcycles the main function is reducing drag when the rider assumes the optimal aerodynamic configuration with his or her body in unison with the machine and does not shield the rider from wind when sitting upright.

Safety

Early windshields were made of ordinary window glass, but that could lead to serious injuries in the event of a crash. A series of crashes led up to the development of stronger windshields. The most notable example of this is the *Pane vs. Ford* case of 1917 that decided against Pane in that he was only injured through reckless driving. They were replaced with windshields made of toughened glass and were fitted in the frame using a rubber or neoprene seal. The hardened glass shattered into many mostly harmless fragments when the windshield broke. These windshields, however, could shatter from a simple stone chip. In 1919, Henry Ford solved the problem of flying debris by using the new French technology of glass laminating. Windshields made using this process were two layers of glass with a cellulose inner layer. This inner layer held the glass together when it fractured. Between 1919 and 1929, Ford ordered the use of laminated glass on all of his vehicles.

Automobile windshield displaying "spiderweb" cracking typical of laminated safety glass.

Modern, glued-in windshields contribute to the vehicle's rigidity, but the main force for innovation has historically been the need to prevent injury from sharp glass fragments. Almost all nations now require windshields to stay in one piece even if broken, except if pierced by a strong force. Properly installed automobile windshields are also essential to safety; along with the roof of the car, they provide protection to the vehicle's occupants in the case of a roll-over accident.

Today's windshields are a safety device just like seatbelts and airbags. The urethane sealant is protected from UV in sunlight by a band of dark dots around the edge of the windshield. The darkened edge transitions to the clear windshield with smaller dots to minimize thermal stress in manufacturing. The same band of darkened dots is often expanded around the rearview mirror to act as a sunshade.

Other Aspects

In many places, laws restrict the use of heavily tinted glass in vehicle windshields; generally, laws specify the maximum level of tint permitted. Some vehicles have noticeably more tint in the uppermost part of the windshield to block sunglare.

In aircraft windshields, an electric current is applied through a conducting layer of tin(IV) oxide to generate heat to prevent icing. A similar system for automobile windshields, introduced on Ford vehicles as "Quickclear" in Europe ("InstaClear" in North America) in the 1980s and through the early 1990s, used this conductive metallic coating applied to the inboard side of the outer layer of glass. Other glass manufacturers utilize a grid of micro-thin wires to conduct the heat especially on the later European Ford Transit vans. These systems are more typically utilized by European auto manufacturers such as Jaguar and Porsche.

The use of thermal glass prevents some navigation systems from functioning correctly, as the embedded metal blocks the satellite signal.The RF signal tends to flow along the metal wires or layer so very little radiation can pass. This can be resolved by using an external antenna. Mobile telephones can also have problems; thermal glass typically allows only 0.0001 (1‰, or one per mille or 1 ppt) of the signal to pass, whereas a concrete wall with rebars allows up to 0.0100 (10%, or 100‰) of the signal to pass.

Terminology

The term windshield is used generally throughout North America. The term windscreen is the usual term in the British Isles and Australasia for all vehicles. In the US windscreen refers to the mesh or foam placed over a microphone to minimize wind noise, while a windshield refers to the front window of a car. In the UK, the terms are reversed, although generally, the foam screen is referred to as a microphone shield, and not a windshield.

Singled aero screen on Bentley Blower No.1.

Sports or racing cars would sometimes have aero screens, which were small semi-circular or rectangular windshields. These were often mounted in pairs behind a foldable flat windshield. Aero screens are usually less than 20 cm (8 in) in height. They are known as aero screens because they only deflect the wind. The twin aeroscreen setup (often called Brooklands) was popular among older sports and modern cars in vintage style.

A *wiperless windshield* is a windshield that uses a mechanism other than wipers to remove snow and rain from the windshield. The concept car Acura TL features a wiperless windshield using a series of jet nozzles in the cowl to blow pressurized air onto the windshield. Also several glass manufacturers have experimented with nano type coatings designed to repel external contaminants with varying degrees of success but to date none of these have made it to commercial applications.

Repair of Stone-chip and Crack Damage

According to the US National Windshield Repair Association, many types of stone damage can be successfully repaired. Whether the windshield can be repaired always depends upon four factors: the size, type, depth and location of the damage.

Size and Depth

Repair of cracks up to 6.1 cm (2.4 in) is within permissible limits; automobile glass with more severe damage needs to be replaced. However, this is dependent on local laws. If a crack extends to the edge of the panel then this would compromise the structural integrity of the windshield. Aircraft windshields are designed in such a way that even if a crack were to extend all the way across the panel, the structural integrity is maintained via multiple failsafe methods in both frame and the glass plies. A sacrificial outer layer that cracks rather than devitrifies is the first failsafe.

Type

Circular bullseyes, linear cracks, crack chips, dings, pits and star-shaped breaks can be repaired without removing the glass, eliminating the risk of leaking or bonding problems sometimes associated with replacement.

Location

Some damages are very difficult to repair, or cannot be repaired:

- On inside of the windshield.
- Deep damage on both layers of glass due to solar absorption or oxidation.
- Damage over rain sensor or internal radio antenna.
- Complex multiple cracks.
- Very long cracks (i.e. over 18–24 inches (46–61 cm) long).
- Contaminated cracks.
- Edge cracks.

In cracked windshield repair, air is removed from the damaged area on the windshield with a specified vacuum injection pump. Then using the injection pump, the clear adhesive resin is injected to replace the air in the windshield crack. The resin is then cured with an ultraviolet light. When done properly, the damaged area's strength is restored, as is 90–95% of the clarity.

Replacement

Windshields that cannot be repaired have to be replaced. Replacement of a windshield typically takes less than an hour. To ensure the vehicle is safe to drive, time values called the Safe Drive Away Time have been established. Windshields which have been replaced must cure or bond sufficiently until they are able to withstand the forces of a crash. Knowing the minimum time needed to cure the glass bonding adhesives is therefore important. This safe drive away time (SDAT) or minimum drive away time (MDAT) refers to the time required until a windshield installation or glass replacement is considered safe to drive again. Criteria are specified in U.S. Federal Motor Vehicle Safety Standards 212/208 to ensure the reliability of adhesive systems. Typically the SDAT is verified with crash tests as well as with high-speed laboratory test methods.

Consumers may be unaware that the MDAT or SDAT time is focused on safety and not necessarily on the quality, durability, or warranty of the installation. Care must be taken not to drive the vehicle prior to the SDAT/MDAT. If a vehicle is released to be driven before the SDAT and the adhesive used to set the new windshield has not had appropriate cure time, the occupants will not be properly protected in the event of a collision.

Airbags deploy at speeds up to 200 mph (320 km/h; 89 m/s) and in some cases exert tremendous force on the windshield. Occupants can impact the airbag just 50 ms after initial deployment. Depending on vehicle design, airbag deployment and occupant impact into the airbag may increase forces on the windshield, dramatically in some cases. Forces of occupants on the airbags - and hence the potential forces on the windshield - are lower for belted occupants. As consequence, adhesive suppliers usually inform their customers about the level of security achieved:

- Example: Security exceeding FMVSS 212/208 belted .

- Example: Security exceeding FMVSS 212/208 unbelted.

With the advent of quick-cure adhesives, mobile windshield replacements have become more prevalent. Often the temperature and humidity cannot be controlled for mobile installations. For most common glass adhesives the ideal environment is 70 °F (21 °C) and 50% humidity. Variations from the ideal curing environment can increase the time needed for a sufficiently safe bond to form. Because of the variables and difficulties involved in mobile windshield replacement, many vehicle manufacturers do not recommend this method of installation.

Architectural Glass

It is difficult to conceive the contemporary architecture without glass. In combination with modern technologies and materials such as steel, concrete, aluminium and other materials, this ancient building material very successfully contributes to extraordinary appearance of buildings. Regardless of it being used for windows, façade or interior partitions, glass connects the space, improves the quality of space, transmits sufficient light, and the contemporary types of glass may contribute to energy saving. It is known that energy saving is one of the most important architectonic challenges of our age. The heat loss through the glass surfacing on the façade or the roof has been significantly reduced owing to modern glass production and processing technologies. Also, more than ever before, there is a concern about the safety of the users and the structure itself. Glass must nowadays conform to the high standards regarding safety of the users and passers-by, thus they are made resistant to shocks and abrupt temperature changes, and in chase they are damaged or shattered, they would not break in. The manufacturers tried in this way to keep the risk of injury to a minimum.

Glass nowadays is an integral part of many facades and roofs. This material can be easily shaped and installed, crating in this way the structures which are gripping and dominating. However, apart from esthetic criteria, a contemporary structure must meet a number of criteria which are necessary for creation of adequate comfort within a structure. In order to improve the comfort of the occupants by an increase in the quality of interior space and optimization of natural resources, it is necessary to conceive of a building with an "interactive" envelope.

When selecting the type and use of glass in a project, one looks for an optimal balance between aesthetics and function. The wide variety of architectural glass commercially available coupled with the versatility and creativity one can explore with the material makes the design process exciting and challenging. The transparency and translucency of glass has historically given an aesthetic quality to architecture like no other material. It gives a building the ability to change, to move, and to create certain environments. The way in which light transmits through a piece of glass in building can be a powerful design tool for an architect. Glass can reflect, bend, transmit, and absorb light, all with great accuracy. Most architectural glass is partially transparent with little reflectance and absorbency. There are hundreds of glass compositions as well as different coatings, colors, thick-nesses, and laminates, all of which affect the way light passes through the material.

Contemporary glass may adapt to variety of architectural forms.

Types of Architectural Glass

The development of the float glass process in the 1950s allowed the economical mass production of high quality flat glass and virtually all architectural glass is now produced by this process. The focus of intensive research and development aimed at maximizing its tree most attractive traits: the ability to transmit light, block heat and safety issues. These efforts have engendered a number of significant advances, from the introduction of uncoated spectrally selective glass to the rise of multi-cavity insulating glass units. Safety issues have a high importance on glass applications, because of potential life safety hazard to pedestrians and building occupants. Today, the vast majority of new windows, curtain walls and skylights for commercial building construction have insulating glazing for energy efficiency and comfort. Nevertheless, high-quality glass products also give the opportunity to design load-bearing structural elements or systems constructed primarily of glass, such as staircases, floors, walls and bridges.

Architectural glass comes in three different strength categories: annealed glass, heat-strengthened glass and fully-tempered glass. Annealed glass is the most commonly used architectural glass. It has good surface flatness because it is not heat-treated and therefore not subject to distortion typically produced during glass tempering. On the downside, annealed glass breaks into sharp, dangerous shards. Heat-strengthened and fully-tempered glass are heat-treated glass products, heated and quenched in such a way to create residual surface compression in the glass. The surface compression gives the glass generally higher resistance to breakage than annealed glass. Heat-strengthened glass has at least twice the strength and resistance to breakage from wind loads or thermal stresses comparing to annealed glass. The necessary heat treatment generally results in some distortion compared to annealed glass. Like annealed glass, heat-strengthened glass can break into large shards. Fully-tempered glass (toughened glass) provides at least four times the strength of annealed glass, which gives it superior resistance to glass breakage. It is float or plate glass that has been heated and rapidly cooled, increasing its inherent strength and ductility. Similar to heat-strengthened glass, the heat-treatment generally results in some distortion. If it breaks, fully-tempered glass breaks into many small fragments, which makes it suitable as safety glazing under certain conditions. It is used for windows that are exposed to high wind pressure or extreme heat or cold. Properties of annealed and fully-tempered glass are comparatively provided in Table.

Table: Properties of Glass.

	Annealed glass	Toughened glass (fully tempered)
Strength	59–150 N/mm²	7–28 N/mm²
Young's modulus	70 kN/mm²	70 kN/mm²
Density	2.4 kg/m³	2.4 kg/m³
Thermal coefficient of expansion	$8.8*10-6$ K⁻¹	$8.8*10-6$ K⁻¹
Poisson's ratio	0.22	0.22

The following are specialized glass types that are made with different qualities to enhance their performance:

- Laminated glass involves sandwiching a transparent sheet of polymer, such as polyvinyl butryal, between two or more layers of flat glass using an adhesive. Because it can prevent the fall-out of dangerous glass shards following fracture, it is often used as safety glazing and as overhead glazing in skylights. It is a durable and versatile glass with plastic interlayer which provides protection from ultraviolet rays and attenuates vibration, and gives laminated glass good acoustical characteristics. Can be used in a variety of environments.

- Insulating glass consists of two or more lites of glass separated by a hermetically sealed space for thermal insulation and condensation control. The airspace between the glass lites can be filled during the manufacturing process with either dry air or a low-conductivity gas, such as sulfur hexafluoride or argon. The thermal performance of double-glazed or triple-glazed windows can be further improved by the addition of a low-emissivity coating on one or all of the layers of glass. The air space also reduces heat gain and loss, as well as sound transmission, which gives the insulating glass superior thermal performance and acoustical characteristics compared to single glazing. Most commercial windows, curtain walls, and skylights contain insulating glass.

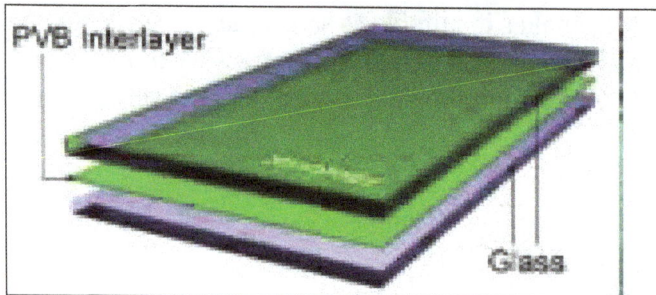

Appearance: a) tempered glass, b) laminated glass.

- Coated glass is covered with reflective or low-emissivity (low-E) coatings. In addition to providing aesthetic appeal, the coatings improve the thermal performance of the glass by reflecting visible light and infrared radiation.

- Tinted glass contains minerals that color the glass uniformly through its thickness and promote absorption of visible light and infrared radiation.

- Wire glass involves steel wires rolled into sheets of glass. A wire mesh is inserted during the manufacturing of plate glass, allowing the glass to adhere together when cracked. It can qualify as safety glass for some applications.

Special Glass Coatings

Glass provides high compression strength and perfect transparency – but also the possibility to alter its transparency through the integration of materials which have a switchable light transmissivity. Today's coating technologies, as well as the possibility of reinforcing glass with different stiffening materials; open a nearly endless range of new ways of using glass. Glass and façade manufacturers now offer a wider range of affordable glazing system solutions which will provide better thermal and solar control without sacrificing daylight, and perhaps control surface temperature at the inside face of the glass to maintain human comfort.

Self-cleaning or easy-to-clean glass uses titanium dioxide coatings as a catalyst to break up organic deposits. It requires direct sunlight to sustain the chemical reaction and rainwater to wash off the residue. Anorganic deposits are not affected by the coatings. Photochromic coatings incorporate organic photochromic dyes to produce selfshading glass. Originally developed for sunglasses, these coatings are self-adjusting to ambient light and reduce visible light transmission through the glass. They provide a more evenly (in terms of time) distributed illumination of interior space regardless of exterior variations and they are typically used to provide shading.

Glass with electrochromic coatings utilizes a small electrical voltage, adjusted with dimmable ballasts, to adjust the shading coefficient and visible light transmission.

Upon switching off the power, they retain the same degree of dimming. In this way it is possible to control the shading of the façade, and thus illumination and temperature of the interior. Like photochromic coatings, they are intended to attain lighting energy savings.

Electrochromic coatings.

Thermochromic laminated glazing (TLG) enables to regulate daylight, automatically adapting dynamically to the continuously changing climatic conditions, aids in reducing the energy needs of a building and providing thermal comfort. Neither electrical power nor driving unit are required. The polymeric interlayer of TLG is doped with complexes of transition metals, which change their coordination and transmission or color of the film under influence of light and heat fluxes.

They are favorable for regulation of interior temperature in comparison to the photochromic glass, because the external temperature and degree of illumination need not be directly mutually dependent, especially in winter.

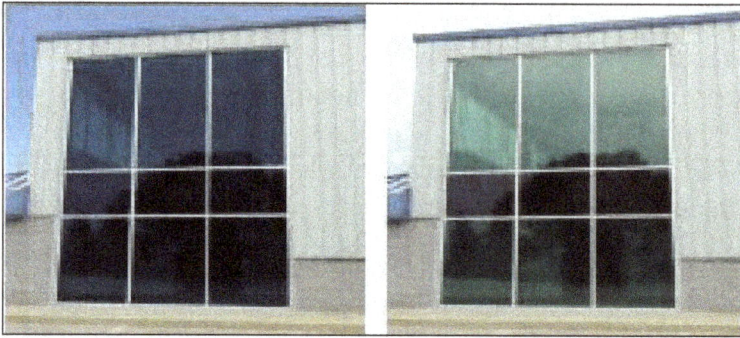

Sunlight responsive thermochromic glass: (a) windows are tinted by direct sunlight; (b) windows are clearer as they are only exposed to indirect sunlight.

Potential for Future Development

To attain, new social, economic and technological ideals architects and engineers of today must improve the quality of buildings and establish new principles of conceptual design of buildings. The quality of interior space and the impact of a building on its surroundings depend strongly on the physical interface that separates the outer environment from the inner building space. The conception and realization of this envelope are therefore of prime importance.

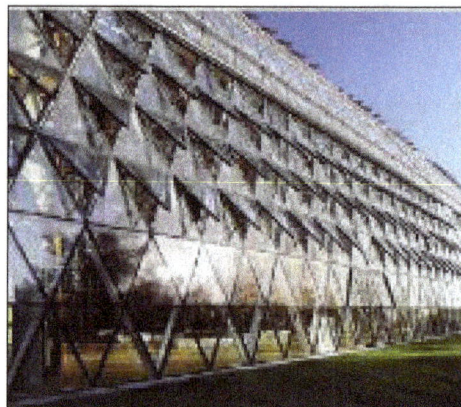

Glass combined with contemporary technical systems for ventilation, shading and collecting of the Sun energy.

Considering market issues it is obvious that growth of demand for glass is not only a consequence of economic growth but also regulations concerning safety, reduction of noise and growing demands in terms of energy efficiency of buildings. However, the lack of standards regulating use of glass in engineering can, in perspective, reduce the growth of glass use in construction.

The main growth drivers that will influence the flat glass consumption in buildings and civil engineering respond to several contemporary demands, presented in Table.

Table: Growth drivers for future flat glass consumption.

Demand	Growth Drivers
Energy saving (heating)	Energy saving legislation and building regulations, reduction of energy loss from buildings and energy labeling of windows.
Energy saving (cooling)	Energy saving legislation, reduction of air-conditioning load in buildings. Preventing non air-conditioned buildings from overheating
Safety	Increasing legislative requirement for safety glass
Security	Requirement for transparency combined with security/safety features
Fire protection	Compliance with fire regulations combined with requirements for good light transmission
Acoustic	Increasing noise levels caused by traffic, aircrafts etc progressively covered by legislation
Self-cleaning	Reduce use of detergents and improve safety of cleaning works on high-rise buildings. Product range now extended to incorporate self-cleaning features

Safety issues occurring in contemporary glass structure are the result of the lack of knowledge of glass properties and other materials forming a composition with it, and designing, production or construction flaws. It is important to consider them so that they could be avoided. The basic measures taken to protect the glass structures are:

- Application of chemically tempered glass with the polycarbonate core, allowing additional safety and durability;

- Application of multi-layered glass, with at least one layer of semi-tempered glass, which will prevent the unwanted displacement of glass panel in case of breakage;

- In case of point supported glass façades, prevention of failure chain reaction by independent support of glass panels in vertical rows, with regular distribution of stress in case of breakage of some of the panels, transferring the load horizontally to adjacent panels;

- In case of point supported glass facades, if possible, provide a central suspension point for the topmost panel in a row, providing an absolutely vertical position;

- In case of multi-layered glass, it is good practice to use one additional layer on the glass than what is required by the static design.

Structural use of glass.

Glass structural design standards are essential to limit the number of structural failures that arise from avoidable negligence resulting from poor design or construction practice due to the lack of a harmonized standard.

The research in the field of enhancing glass as a building material, its properties and processing methods has resulted in a range of various products, which is huge stimulation to expand its use in architecture and beyond. Technical issues for the growing application of glass in civil engineering require to establish and harmonize certain criteria of its production, processing and installation. The upgrade of the current situation, where different standards are in use in different countries, can be achieved by the harmonization and development of the single European standard.

Table: Main issues related to the further development of standards to help the practical application of the knowledge gained in the application of glass as structural products.

Item / Issue	Level of solution /perception
Availability of materials	High
Versatility of products	High
Technology level	High
Cost	Moderate
Codes, standards and specifications	good, but still insufficient: lack of harmonized design standards
Supportive data banks	Fair
Understanding of behavior	weak, perceived unreliability
Knowledge of structural mechanics	relatively high
Knowledge of design methodology	Insufficient
Knowledge of detailing	Poor
Confidence in reliability of structure	Low

Current Status of Standards and Regulations

Regulations are the result of experience and research and should undergo updating in order to create an adequate framework for designing, construction and production. When it comes to civil engineering, regulations are very extensive and difficult to consider. In most cases, it is necessary to consult several reference documents in order to fully understand the observed problem. As for the architectural glass, it is important to mention that there are several leading international standards such as the U.S standard E1300-07, British BS 6262 and European pre-standard prEN 13474. The American standard is recognized in the countries which do not have their own glass standards; the British standard is relevant for the British and Irish market where construction of glass-aluminium facades is very prominent, while the European pre-standard is an attempt to harmonize the design of glass structures for the entire European market. This would facilitate designing, application and monitoring of architectural glass. The globalization of the construction market comprising construction products, engineering and construction services requires international standards families in order to avoid inconsistencies due to the use of various national standards.

Each of the mentioned standards is composed of a set of design standards in connection with product and testing standards.

In the EU regulations, three groups of regulations can be singled out, which must be considered when designing elements and structures of glass:

- General standards, related to designing and construction in general,
- Standards relating to glass as a building material,
- Standards relating to glass structures.

Particular attention must be paid to the last group of standards relating to: safety and security of the structure, fire safety, acoustic comfort, thermal comfort, water vapor diffusion, water and air permeability.

The national standards in this area are fairly poor and based on the international standards. As there is a multitude of structural elements of glass, it is necessary to define and structure this field by a number of standards and regulations. The structural application of glass and static designs for glass facades and structural safety of glass buildings are not the subject of national standards.

In Germany, one of the leading countries in this field, there is a number of standards and rules, as a DIN 18008 which provides rules for design and construction as well as specifications for required experimental verifications. In Germany, technical rules for glass construction are available and include technical rules for glazing acting as anti-drop device/railing (TRAV), technical rules for linear supported glazing (TRLV) and technical rules for point supported glazing (TRPV).

Fire-proofing is a very important characteristic when it comes to contemporary materials. Two most important European standards dealing with these issues are: DIN 4102 and ISO 834-10. In national regulations, SRPS ISO 3009:1993 regulates fire-proof characteristics of glazed elements.

Acoustic comfort, as a part of engineering physics, represents the youngest field of research as opposed to other – classical topics in civil engineering. The acoustic comfort is defined by the standard SRPS EN ISO 140 and SRPS EN ISO 717, dealing with laboratory measuring of insulation capacity R and determination of the level of sound-proofing in buildings. The national standard SRPS EN 12354-1-6:2000 defines the level of sound-proofing of buildings on the basis of acoustic performances of building elements.

Thermal comfort is a developing area and it is very interesting. In the field of thermal protection, there are the following standards: EN 13790-thermal performance of the structure, EN 13830-thermal aspect of curtain walls, EN ISO 10077-1-includes the method for design of thermal characteristics of thermal insulation glazing, EN 1279 1-6 regulates the field of thermal insulation glass. In the national regulations the current standard is SRPS EN 1279-1-6 related to construction glass, which is taken from and partly translated European standard EN 1279 1-6. The standards treating the methods of calculation of thermal transmission coefficient (U value) are SRPS EN 673 - Determination of thermal transmittance-calculated method, SRPS EN ISO 10077 1-2 - Thermal performance of windows, doors, shutters - Calculation of thermal transmittance. Since 2008 in the national standards there has been the regulation SRPS EN 13947 defining calculation of thermal transmittance of curtain walls. The calculation includes: different types of glazing, e.g. glass or plastic; single or multiple glazing; with or without low emissivity coating; with cavities filled with air or other gases; frames (of any material) with or without thermal breaks; different types of opaque panels clad with metal, glass, ceramics or any other material. Thermal bridge effects at the rebate or connection between the glazed area, the frame area and the panel area are included in the calculation.

Water vapor diffusion, represents an important characteristic when the external partitions are concerned, whereas, in national standards it relates only to massive structures. The European standard EN ISO 13788:2001 deals with this problem and calculation methods.

Since recently, the field of glass facades was arranged in terms of permeability to air and water due to variations in pressure. The standards covering this field are SRPS EN 12152, 12153, 12154 and they specify requirements and classification of air permeability of both fixed and openable parts of curtain walling, also defining the method to be used to determine the air permeability and defining the requirements and classification of water tightness performance of curtain walling.

References

- Contact-lenses, conditions-treatments: umkelloggeye.org, Retrieved 10 March, 2019

- Advantages-and-disadvantages-of-various-types-of-contact-lenses, caring-for-your-vision-contact-lenses: aoa.org, Retrieved 08 June, 2019

- Ferreira, J P J G; Branco, F A B (2007). "The Use of Glass Fiber-Reinforced Concrete as a Structural Material". Experimental Techniques. 31 (May/June 2007): 64–73. doi:10.1111/j.1747-1567.2007.00153.x

- Rahaman, M (2011). "Bioactive glass in tissue engineering". Acta Biomaterialia. 7 (6): 2355–2373. doi:10.1016/j.actbio.2011.03.016. PMC 3085647. PMID 21421084

- Rabiee, S.M.; Nazparvar, N.; Azizian, M.; Vashaee, D.; Tayebi, L. (July 2015). "Effect of ion substitution on properties of bioactive glasses: A review". Ceramics International. 41 (6): 7241–7251. doi:10.1016/j.ceramint.2015.02.140.

- Architectural-glass-Types-performance-and-legislation- 274829376: researchgate.net, Retrieved 15 April, 2019

- R Wananuruksawong et al 2011 IOP Conf. Ser.: Mater. Sci. Eng. 18 192010 doi:10.1088/1757-899X/18/19/192010 Fabrication of Silicon Nitride Dental Core Ceramics with Borosilicate Veneering material

PERMISSIONS

All chapters in this book are published with permission under the Creative Commons Attribution Share Alike License or equivalent. Every chapter published in this book has been scrutinized by our experts. Their significance has been extensively debated. The topics covered herein carry significant information for a comprehensive understanding. They may even be implemented as practical applications or may be referred to as a beginning point for further studies.

We would like to thank the editorial team for lending their expertise to make the book truly unique. They have played a crucial role in the development of this book. Without their invaluable contributions this book wouldn't have been possible. They have made vital efforts to compile up to date information on the varied aspects of this subject to make this book a valuable addition to the collection of many professionals and students.

This book was conceptualized with the vision of imparting up-to-date and integrated information in this field. To ensure the same, a matchless editorial board was set up. Every individual on the board went through rigorous rounds of assessment to prove their worth. After which they invested a large part of their time researching and compiling the most relevant data for our readers.

The editorial board has been involved in producing this book since its inception. They have spent rigorous hours researching and exploring the diverse topics which have resulted in the successful publishing of this book. They have passed on their knowledge of decades through this book. To expedite this challenging task, the publisher supported the team at every step. A small team of assistant editors was also appointed to further simplify the editing procedure and attain best results for the readers.

Apart from the editorial board, the designing team has also invested a significant amount of their time in understanding the subject and creating the most relevant covers. They scrutinized every image to scout for the most suitable representation of the subject and create an appropriate cover for the book.

The publishing team has been an ardent support to the editorial, designing and production team. Their endless efforts to recruit the best for this project, has resulted in the accomplishment of this book. They are a veteran in the field of academics and their pool of knowledge is as vast as their experience in printing. Their expertise and guidance has proved useful at every step. Their uncompromising quality standards have made this book an exceptional effort. Their encouragement from time to time has been an inspiration for everyone.

The publisher and the editorial board hope that this book will prove to be a valuable piece of knowledge for students, practitioners and scholars across the globe.

INDEX

A

Acidic Extraction, 38

Annealing, 12, 28, 70, 78-79, 89, 93, 100-101, 106, 116-117, 121-122, 153, 158

B

Benthoscope, 25

Beveled Glass, 46-47

Bioactive Glass, 123, 149-151, 155-156, 200, 219

Bioglass, 123, 149-158, 200

Borosilicate Nanoparticles, 140

C

Caneworking, 90, 94, 108, 201

Cer-Vit, 37, 200

Ceramization, 35

Chemical Etching, 29

Commercial Glass Composition, 2

Controlled Pore Glass, 38, 40

Crystalline Metals, 182

Crystalline Phase, 29, 35, 75

D

Devitrification, 3, 117, 199

Dichroic Glass, 114-115

Dielectric Constant, 10, 14-15, 25, 28

E

Electrochromic Devices, 19

Electronic Conduction, 15

F

Favrile Glass, 56-57

Fibreglass, 6, 54

Foturan, 32-36, 69

Fourcault Process, 54, 118-121

Fracture Glass, 57-58

Free-blowing, 91-92

Fused Quartz, 5, 18, 23-27, 95, 128, 134, 136

Fused Silica, 2-3, 23-27

G

Glass Batch Calculation, 70, 72

Glass Beadmaking, 112-114

Glass Casting, 90, 104

Glass Fiber Reinforced Concrete, 158

Glass Melting Enthalpy, 75

Glass-ceramic, 29, 32, 35, 37, 155, 158, 199-200

Glassblowing, 48, 90-93, 108-111, 115, 137, 140, 200-201

Glassy Metals, 7

Gob Preform, 98

Google Glass, 145, 147

H

Hebron Glass, 18, 47-49

Hotforming, 96

Hydrofluoric Acid, 2, 26, 29, 34-35, 67

I

Index Drop, 100-101

Industrial Glass, 1, 4

K

Kiln Casting, 104-106

L

Lampworking, 114-115, 121-122, 137, 140

Lead-alkali Silicate, 6, 10

Liquid-crystal Devices, 19-20

Liquid-crystal Display, 160

Lithography, 21, 34

M

Megapascals, 11, 43

Millefiori, 108, 110, 113

Mold-blowing, 91-92

N

Nanoparticles, 31, 140

Neoprene Seal, 206

Non Oxide Glasses, 7

Nonsilica, 7

P

Phosphorescence, 26-27

Photosensitive Glass, 29-32

Photosensitivity, 16, 33

Plate Glass, 18, 42, 54, 181, 211, 213

Polarizing Filter, 42

Porous Glass, 37-40

Precision Glass Moulding, 95-99, 101, 103

Preforms, 96-98

R

Refractive Index, 3, 7, 10, 16-17, 28, 31, 33, 55, 74, 84, 86, 100-101, 132, 138, 181, 183

S

Sand Casting, 105

Sandblasting, 202, 204

Sintering Glass Powder, 37

Smart Glass, 18-19, 21-23, 53

Softening Point, 10, 12, 28, 79, 95, 111, 134

Spectral Transmittance, 34

Stained Glass, 1, 18, 47, 56, 59-68, 200-201

T

Temperature Sensitive Glass, 41-42

Tempered Glass, 2, 42-46, 52, 211-212, 215

Thermal Stability, 25

Thermochromic Laminated Glazing (tlg), 214

Tiffany Glass, 18, 55

Torino Scale, 37

U

Ultratransparent Oxide Glasses, 3

V

Vitreous Silica, 5-7, 10, 12-15

Vitrification, 90, 106-108, 122, 139

Y

Young's Modulus, 10, 28, 158, 211

www.ingramcontent.com/pod-product-compliance
Lightning Source LLC
Chambersburg PA
CBHW061952190326
41458CB00009B/2855